Ecological Integrity, Law and Governance

Ecological integrity is concerned with protecting the planet in a holistic way, while respecting ethics and human rights. Over recent years it has been introduced directly and indirectly in several legal regimes, culminating in international law with the 2016 expanded remit of the International Criminal Court, which now includes 'environmental disasters'.

This book celebrates the 25th anniversary of the Global Ecological Integrity Group (GEIG), which includes more than 250 scholars and independent researchers worldwide, from diverse disciplines, including ecology, biology, philosophy, epidemiology, public health, ecological economics and international law. It reviews the role of ecological integrity across a number of fields through inter- and trans-disciplinary engagement on matters affecting and governing the sustainability of life for both present and future generations. These include ethics, environmental disasters, crimes against humanity and environmental health, and how such issues can be subject to sound governance and be incorporated into international law. The book also looks forward to new applications of the concept of ecological integrity, such as crimes that result in the exploitation of natural resources and the illegal dispossession of land.

Laura Westra is Professor Emerita (Philosophy) and Sessional Instructor, Faculty of Law, University of Windsor, Canada, and Visiting Professor, Faculty of Jurisprudence, University of Parma, Italy.

Klaus Bosselmann is Professor of Law and Director of the New Zealand Centre for Environmental Law at the University of Auckland.

Janice Gray is Senior Lecturer in the Faculty of Law at the University of New South Wales, Australia.

Kathryn Gwiazdon is Executive Director of the Center for Environmental Ethics and Law, a US-based non-profit organization.

Ecological Integrity, Law and Governance

Edited by

Laura Westra, Klaus Bosselmann, Janice Gray and Kathryn Gwiazdon

Routledge
Taylor & Francis Group
LONDON AND NEW YORK

earthscan
from Routledge

First published 2018 by Routledge

2 Park Square, Milton Park, Abingdon, Oxfordshire OX14 4RN
52 Vanderbilt Avenue, New York, NY 10017

Routledge is an imprint of the Taylor & Francis Group, an informa business

First issued in paperback 2020

British Library Cataloguing-in-Publication Data
A catalogue record for this book is available from the British Library

Library of Congress Cataloging-in-Publication Data
A catalog record has been requested for this book

ISBN: 978-0-8153-9463-1 (hbk)
ISBN: 978-0-367-51090-9 (pbk)

Typeset in Goudy Old Style by
Servis Filmsetting Ltd, Stockport, Cheshire

Visit the [companion website/eResources]:
[insert comp website/eResources URL]

To the Global Ecological Integrity Group past, present and future

Contents

Contributors

Donato Bergandi, PhD
Country of affiliation: France
Professor
Muséum national d'histoire naturelle
www.environmentalethicsandlaw.org/advisors
bergandi@mnhn.fr

Klaus Bosselmann, PhD
Countries of affiliation: New Zealand, Germany
Professor of Law
New Zealand Centre for Environmental Law
University of Auckland
https://unidirectory.auckland.ac.nz/people/profile/k-bosselmann
k.bosselmann@auckland.ac.nz

Donald A. Brown, JD, MA (Liberal Studies, Philosophy and Art)
Country of affiliation: USA
Scholar in Residence and Professor
Widener University Commonwealth Law School
Dabrown57@gmail.com Ethicsandclimate.org

Martin J. Bunch, PhD
Country of affiliation: Canada
Professor and Associate Dean (Research)
Faculty of Environmental Studies, York University
bunchmj@yorku.ca
www.yorku.ca/bunchmj

Colin D. Butler, PhD, BMed
Country of affiliation: Australia
Adjunct Professor
University of Canberra
Australian National University

Flinders University
https://colindbutler.weebly.com/ colin.butler1955@gmail.com

Paul Carrick, MA, PhD (Philosophy)
Country of affiliation: USA
Professor Emeritus
Gettysburg College
Kienle Center for Humanistic Medicine of Pennsylvania State University
Harrisburg Area Community College
pcarrick@gettysburg.edu

Sheila D. Collins, PhD
Country of affiliation: USA
Professor Emerita (Political Science)
William Paterson University
sheila.collins65@verizon.net

Philippe Crabbé, Docteur en Droit, Licencié en science économique
Country of affiliation: Canada
Professor Emeritus
University of Ottawa
crabbe@uottawa.ca

Joseph W. Dellapenna, BBA, JD
LLM (International and Comparative Law)
LLM (Environmental Law)
Country of affiliation: USA
Attorney at Law
Independent
dellapen@law.villanova.edu

Rose A. Dyson, EdD
Country of affiliation: Canada
Media Education Consultant
President
Canadians Concerned about Violence in Entertainment
www.C-CAVE.com
rose.dyson@alumni.utoronto.ca

J. Ronald Engel, PhD
Country of affiliation: USA
Professor Emeritus
Meadville Lombard Theological School
jronengel@gmail.com

Joan Gibb Engel, PhD (English, Creative Writing)
Country of affiliation: USA
jgengel22@gmail.com

Michelle Gallant, PhD (London), LLM (UBC), LLB (UNB), BA (UPEI)
Country of affiliation: Canada
Professor
University of Manitoba
gallantm@umanitoba.ca

Geoffrey Garver, PhD, LLM, JD, BS
Country of affiliation: Canada
Adjunct Professor/Sessional Lecturer
McGill University and Concordia University
gginmont@videotron.ca

Marco Ettore Grasso
Country of affiliation: Italy
PhD (Philosophy, Sociology and Law)
University of Milano Bicocca
marco.grasso2@unimib.it
marcoeg@hotmail.it

Janice Gray, BA, LLB, Grad Dip Ed, MA (UNSW)
Grad Dip Leg Prac (UTS), PhD (Law, UEA)
Country of affiliation: Australia
Doctor
Senior Lecturer
University of New South Wales, Sydney, Australia
j.gray@unsw.edu.au

Kathryn Anne Gwiazdon, JD
Country of affiliation: USA
Executive Director
Center for Environmental Ethics and Law
www.environmentalethicsandlaw.org

Yuliya Lyamzina, PhD, MBA
Country of affiliation: Japan
Assistant Professor
Radiation Medical Science Center for the Fukushima Health Management Survey, Fukushima Medical University
lyamzina@fmu.ac.jp, http://fmu-global.jp/yuliya-lyamzina-phd-mba

Kathleen Mahoney, QC, FRSC
Country of affiliation: Canada
Professor
University of Calgary
kmahoney@ucalgary.ca

Margot W. Parkes, MBChB, MAS, PhD
Countries of affiliation: Canada, New Zealand
Canada Research Chair in Health, Ecosystems and Society, Associate Professor
University of Northern British Columbia
margot.parkes@unbc.ca,
www.unbc.ca/parkes

Robert Rattle, BSc (Physics)
Country of affiliation: Canada
Executive Director
Crane Institute for Sustainability
http://craneinstitute4sus.wixsite.com/ci4s

Colin L. Soskolne, PhD
Countries of affiliation: Canada, Australia
Professor Emeritus, Adjunct Professor
University of Alberta; University of Canberra
Chair
Development Committee, IJPC-SE
www.colinsoskolne.com; www.ijpc-se.org
colin.soskolne@ualberta.ca

Prue Taylor, LLB, LLM (Hons), LLM (Energy and Environmental Law)
Country of affiliation: New Zealand
Senior Lecturer
University of Auckland
Prue.taylor@auckland.ac.nz

Laura Tomie (BHSc, in progress)
Country of affiliation: Canada
lauratomie@hotmail.com

Anne Venton, BA, BEd, MA (Philosophy – University of Toronto)
Country of affiliation: Canada
Alumni Association Executive
Ontario Institute for Studies in Education
Executive
Canadians Concerned about Violence in Entertainment
anne.venton@sympatico.ca

Peter Venton, BA (Economics), MA (Economics)
Country of affiliation: Canada
President
JPV Associates: Economics & Public Policy
Treasurer
Canadian Peace Research Association
Treasurer
Canadian Pugwash Group
peter.venton@bell.net

Laura Westra, PhD, PhD (Law)
Countries of affiliation: Canada, Italy
Professor Emerita (Philosophy)
Sessional Instructor
Faculty of Law, University of Windsor
www.ecointegrity.net
www.globalecointegrity.net

Introduction

The development of the Global Ecological Integrity Group's research and its future aims

Laura Westra and Klaus Bosselmann

As we consider the collection that represents the work of the Global Ecological Integrity Group on their 25th Anniversary, we can be assured that the future work of the group is in excellent hands. The group has developed from researchers faced with a somewhat new and important concept (undefined), but newly entrenched in binational law (in the Great Lakes Water Quality agreement, 1987, rat. 1988), following its presence in the 1971 US Water Quality Act, to researchers and scholars working in multiple disciplines and representing varied values, who now make up the group's most active members.

Before the year 2000, we met on occasion, at first with the meetings of the Great Lakes Association. Eventually the group expanded to include political scientists and ethicists. We had understood ecological integrity as an 'umbrella' concept, one that included many meanings and diverse aspects. We decided that the best starting point for the first phase of our research would be to reach a collaborative definition that encapsulated most of integrity's diverse aspects.

Hence philosophers, ecologists, biologists, complex system theorists and other scientists found a way to cooperate, and after much discussion and many re-writes, eventually succeeded in reaching a definition that we all could sign our name to comfortably. This definition was published in what became our first collection, *Ecological Integrity: Integrating Environment Conservation and Health* (edited by David Pimentel, Laura Westra and Reed Noss; Island Press, 2000; see also further information at www.globalecointegrity.org). Many more have been published since that time.

That was also the year when we had our first meeting in Costa Rica, after our first of two meetings with the WHO 1999 (first in Rome, Italy, then later in their offices in Budapest, Hungary). In Costa Rica I (Laura) met with Steven Rockefeller, who was working on the final draft of the Earth Charter, and who promised to include me in the next meeting of that enterprise at his home in New York State. The Earth Charter is a particularly important document from our point of view, because it is the only such ethical/political instrument that names ecological integrity as a major principle to guide public policy, something I had advocated in my books in the 1990s (*The Principle of Integrity*, Rowman & Littlefield, 1994; *Living in Integrity*, Rowman & Littlefield, 1998). The Earth

Charter identified Ecological Integrity as its second principle, after 'respect and care for the community of life' (see www.earthcharter.org).

After 2000, many new players joined the group, and a new discipline dominated our research: the law. In fact I also abandoned teaching philosophy and entered law school, where I received my second PhD in 2005. It was obvious that despite the most carefully crafted, authoritative and convincing list of principles counselling policies based on justice, fairness and respect for both nature and human dignity, the necessary policy changes simply did not materialize. Hence it seems that exhortations to morality, and the suggestions that followed, were simply not efficacious enough, so that the law appeared to be a better option for the achievement of environmental and human justice.

The development of the work of the Global Ecological Integrity Group can be understood as involving three distinct phases: first, a phase of clarification (culmination in the 2000 definition); second, a phase of dissemination (spreading the work to new disciplines); and third, the present phase of implementation. The clarification period focused mainly on achieving a science-based definition, because others were keen to introduce a social aspect to the understanding of integrity, which tended to render the concept too vague and subjective to be used as a foundational principle. Only a solid scientific definition could be used to base subsequent social justice norms and, as mentioned above, that definition was reached in 2000.

In the second phase, we worked to demonstrate the relevance of ecointegrity to most disciplines and realms of knowledge, starting with public health, which was and still remains the closest to the science of integrity and the best connection between integrity and human rights. The final period uses the findings of both previous periods to promote ecointegrity through law and global governance: and that is the main aspect to be found in the chapters of the present collections.

At this time, our main effort is to explore a number of new avenues to achieve the serious consideration of ecological integrity in a number of fields. Most of these fields are well represented in the numerous works the group's members have authored through the years, singly and collectively, as well as in the present collection. At any rate, the 'implementation stage' does not exclude theoretical considerations, as demonstrated by Philippe Crabbé in his chapter on early philosophical considerations on ecology. The other chapters in Part I address a wide range of policy issues: Katy Gwiazdon discusses in detail a recent and ongoing case (US) dealing with climate change, while Rose Dyson analyses the difficulties that the 'digital age' adds to the disaster that climate change represents. Marco Ettore Grasso adds a further critical discussion of the problem, and Michelle Gallant exposes the further difficulties imposed by policies of secrecy regarding global taxation issues, which contribute to the problems that exist to reach ecologically sound policy choices.

Part II demonstrates the interface between the protection of ecological integrity and public health, as both are based on human rights, inseparable as they are from natures rights. Kathleen Mahoney addresses Indigenous human rights, explaining the success of her case against the government of Canadas early actions and their treatment of the First Nations. It is worth recalling that the defence of wild areas and their integrity is basic to First Nations' rights in Canada. Joseph Dellapenna discusses a 2017 case, that of the Dakota pipeline. Colin Soskolne et al. discuss the important emergence of the concepts of 'ecohealth' or 'one health', which combine public health and ecological concerns. Robert Rattle and Laura Tomie expose the public health impacts of the Trans-Pacific Partnership presently under consideration by many countries, including Canada. Yuliya Lyamzina discusses the grave consequences of a recent environmental disaster in Japan, the Fukushima nuclear accident. Anne Venton's chapter reprises the important theme of environmental disasters in a case study involving nuclear power in southern Ontario. Finally, Janice Gray exposes the 'causes for concern' related to the increasing spread of unconventional gas mining.

Part III addresses some major problems in governance, starting with Sheila Collins's detailed discussion of environmental problems during Donald Trump's US presidency. Geoffrey Garver and Donald Brown consider the present grave environmental concerns, from the standpoint of current laws for the former, and the lack of an ethical approach in decision-making, for the latter, while Prue Taylor outlines the serious environmental problems that affect our oceans.

The final part, Part IV, looks forward in various ways, some entirely unconventional. Donato Bergandi and Paul Carrick offer illuminating discussions of the major problems we have seen and the philosophical position that would best help to address them. Peter Venton suggests the approach of global governance based on world federalism, while Ron Engel proposes a 'renewal' and reconsideration of the Earth Charter as a starting point. Joan Engel proposes an approach through art, specifically through poetry, as that might offer a novel way of influencing climate change policy. In his concluding chapter Klaus Bosselmann shows that ecological integrity is now being increasingly acknowledged in both law and governance.

Part I

Ecological integrity, ethics and the law

1 The state versus the environment

The ethical and legal implications for state non-action in protecting the foundations of life

Kathryn A. Gwiazdon

Introduction

> He has plundered our seas, ravaged our coasts, burnt our towns, and destroyed the lives of our people.
>
> (United States Declaration of Independence, 1776)

A social contract unites every citizen with its state: individuals, noting the lack of certain protections if they acted merely as individuals, form a community in order to better preserve their natural rights, and in so doing, give up certain individual freedoms. The most influential exploration of social contact theory is John Locke in *Two Treatises of Government* (1698), noting that 'individuals willing to join in society with others, who are already united, or have a mind to unite, [do so] for the mutual preservation of their lives, liberties and estate', yet acknowledging that 'when they enter into society, give up the equality, liberty, and executive power they had in the state of nature, into the hands of the society, to be so far disposed of by the legislative, as the good of the society shall require'.[1]

From this understanding of social contract theory, which has been laid down in state and national constitutions around the world, the primary duty of the state is to protect. A state must protect its citizens from threats and violence that they could not face as individuals. A state must provide goods, services, and an infrastructure that allows for economic, social, and public security. A state must invest in public health, environmental protection, and education to allow citizens to be active, healthy, and informed members of civil society. This article will argue that the key tenet of the state's duty to protect is that it protect the foundations of life upon which its citizenry, and the rights of its citizenry, depends.

Locke argued that it was the government's obligation to protect the life, liberty, and property of its citizens, 'And all this to be directed to no other end, but the peace, safety, and public good of the people'.[2] But what if the state does not uphold the social contract made between it and the citizen? What if the state not only fails to create, implement, or enforce laws that allow for the citizen to exercise its rights, but what if it openly assaults existing laws? The notion of the 'state versus the environment' explores the inaction and sometimes violent aggression of states against the public welfare, against the foundations of life (i.e. soil, water, air, ecological systems integrity, biodiversity, etc.), and argues that this not only violates individual rights, but also undermines the strength and stability of the state itself.

Historically, and in the most extreme instances, when a state breaks its social contract, it leaves the citizens no longer bound to the state; thereby leading to uprisings and revolutions. According to Locke, dissolve the government and re-create it, 'as they shall find it most for their safety and good'.[3] But it shouldn't have to come to that. This article seeks to explore the fundamental duty of the state to protect, and some of the avenues that citizens have taken when the state fails to do so. It will begin by discussing the responsibilities of the state, as the protector, the provider, and the investor. The article will then explore the ethical and legal implications for a state's failure to act to protect the foundations of life by exploring the global trend of climate change litigation brought by children against their states. This will be done primarily by unpacking a federal climate change case in the United States, *Kelsey Cascadia Rose Juliana, et al., v. United States of America, et al.* (*Juliana*).

To further elucidate this trend, the article will briefly highlight recent cases from Ireland, India, New Zealand, Netherlands, Pakistan, and the European Court of Human Rights, as well as developments with the International Criminal Court, the UN Human Rights Council, and the Declaration on the Rights of the Child. It will conclude that in protecting a citizen's fundamental rights, it is the state's most imperative, most basic, ethical, and legal duty to protect that which all rights depend, upon which all citizens depend: the foundations of life. For, '[i]f the people cannot trust their government to do the job for which it exists – to protect them and to promote their common welfare – all else is lost'.[4]

The responsibilities of the state

> The first duty of a Government is protection. The crowning glory of a Republic is that it leaves no man, however humble, without protection. Show me a man exposed to wrong, and I show you an occasion for the exercise of all the power that God and the Constitution have given you.[5]

The Honorable Charles Sumner of Massachusetts spoke these words before Congress in 1867. Sumner believed that President Johnson put an entire nation at risk by advancing values contrary to freedom and equality, contrary to the state's role as the protector of all its citizens. President Johnson ascended to the US Presidency after the assassination of President Abraham Lincoln in 1865. During the Reconstruction Era following the Civil War, after the assassination of President Lincoln, he sought to grant broad amnesty to Confederate leaders and deny civil rights to former slaves; he opposed the Civil Rights Act of 1866, the Thirteenth Amendment, abolishing slavery, the Fourteenth Amendment, affording equal protection and due process by states, and the Fifteenth Amendment, prohibiting voter disenfranchisement based on race, colour, or previous condition of servitude. The Fifth, and later the Fourteenth Amendment, carried the spirit and text of Locke into the U.S. Constitution: the prohibition of federal, state, and local government officials from depriving persons of life, liberty, or property without procedural and substantive due process.

Many saw President Johnson's actions as an affront to the values of America and even succeeded in impeaching him.[6] Federalist Paper No. 57, attributed to Alexander Hamilton or James Madison, stated:

> The aim of every political constitution is, or ought to be, first to obtain for rulers men who possess most wisdom to discern, and most virtue to pursue, the common good of the society; and in the next place, to take the most effectual precautions for keeping them virtuous whilst they continue to hold their public trust.[7]

President Johnson, the chief magistrate of the state, unsuccessfully tried to deny the state's duty to protect all, a value held so dear to the ideals of the nation that it had caused a civil war.

Thomas Hobbes stated in *Leviathan*, 'He that is to govern a whole Nation, must read in himself, not this, or that particular man; but Man-kind.'[8] The state is the protector of the whole: the people, the land, the waters, the air. As the protector of the whole, it must not only protect its citizens from internal and external threats, but also guarantee that the 'whole' is well cared for, providing the best foundations for which its citizens can flourish and exercise their fundamental rights.

As a protector, the state must provide: it must provide goods and services that individuals cannot provide for themselves, such as a healthy environment, roads, water, sanitation, etc. This role as a provider also includes providing an infrastructure of care that allows citizens to flourish economically and socially; for example, drafting and upholding laws that benefit the common good and ensure reliability and stability, and providing institutional checks on corruption and abuses so that the citizenry has confidence in the state. And as a protector, the state must also invest. It must invest in the environment, health care, and education to allow citizens to be healthy, resilient, and ultimately, better provide for themselves.

In the 2015 'Laudato Si' on Care for our Common Home', Pope Francis wrote, 'Climate change is a global problem with grave implications: environmental, social, economic, political and for the distribution of goods. It represents one of the principal challenges facing humanity in our day.'[9] A 2016 poll by the Pew Research Center concluded that, 'Europeans see ISIS, climate change as most serious threats' to their country.[10] And every single nation on Earth has signed the Paris Agreement, except the United States.[11]

If part of the social contract between a state and its citizens is that it will create and enforce laws to protect its citizens, and ensure the conditions for life, liberty, and the pursuit of happiness, what does it mean if a state removes itself from a global pact that seeks to protect life? What does it mean if a state fails to draft environmental laws, or weakens or does not enforce existing environmental laws? What does it mean if the state silences or denies science on climate change or toxins, or prevents scientists from reporting their research to the public? What does it mean if a state does not protect the foundations of life of its citizenry, on which their constitutionally-guaranteed rights rely? Can a state still claim to protect life, when advocating against life?

The ethical and legal implications for failing to protect the foundations of life

Kelsey Cascadia Rose Juliana, et al., v. United States of America, et al: 'This is no ordinary lawsuit'[12]

Juliana effectively alleged a breach of the social contract, as adopted by the U.S. Constitution, between the state and its citizenry: that the government's collective action and inaction on climate change threatens a stable climate system, which directly threatens the Plaintiffs' and future generations' fundamental rights to life and liberty.

Background

On 12 August 2015, 21 youths from across the United States and a former NASA climate scientist, Dr. James Hansen, as 'Guardian of Future Generations', sued the federal government alleging that the government's 'actions and inactions – whether or not they violate any specific statutory duty – have so profoundly damaged our home planet that they threaten plaintiffs' fundamental constitutional rights to life and liberty'.[13] The lawsuit asked whether the government has a constitutional responsibility to leave a viable climate system for future generations, arguing that, 'The federal government has consciously chosen to endanger young people's right to a stable climate system for the short-term economic interests of a few. This administration must no longer consign future generations to an uninhabitable planet.'[14] 'No less important than in the civil rights cases', Plaintiffs seek a court order that would require the President to immediately implement a national climate change action plan that would decrease greenhouse gas (GHG) emissions to safe levels.

Several industries intervened, claiming a 'direct threat to their businesses'.[15] These included American Fuel and Petrochemical Manufacturers (AFPM), representing virtually all refiners and petrochemical manufacturers in the US; the American Petroleum Institute (API), representing 625 oil and natural gas companies; and the National Association of Manufacturers (NAM). Following the 10 November 2016 Opinion and Order, each of these intervenors petitioned, and was allowed, to withdraw from the suit.[16]

Meeting the children: 'My family and I feel very vulnerable'[17]

When the plaintiffs addressed the legal duties for government action to protect the foundations of life, they also addressed ethical duties, such as the duty to prevent irreversible damage, the duty not to harm, the rights of future generations, the precautionary principle, and the duty to protect vulnerable populations.[18]

Lead Plaintiff Kelsey Juliana highlighted the urgency to take action to prevent irreversible destruction, as well as putting an entire generation at risk:

> We brought this case because the government needs to immediately and aggressively reduce carbon emissions and stop promoting fossil fuels, which force our nation's climate system toward irreversible impacts. If the government continues to delay urgent annual emissions reductions, my generation's wellbeing will be inexcusably put at risk.[19]

Xiuhtezcatl Tonatiuh Martinez argued that the government failed to take action to prevent known harms:

> The Federal Government has known for decades that CO_2 pollution from burning fossil fuels was causing global warming and dangerous climate change. It also knew that continuing to burn fossil fuels would destabilize our climate system, significantly harming my generation and generations to come. Despite knowing these dangers, Defendants did nothing to prevent this harm. In fact, my Government increased the concentration of CO_2 in the atmosphere to levels it knew were unsafe.[20]

On 7 February 2017, the case was amended to name President Donald Trump as the defendant. Plaintiff Kiran Oommen took this opportunity to raise the issues of abuse of power, political corruption, and ineptitude, arguing that the president personally benefits from the harms caused and does not understand climate science. He also raised the constitutional obligation of the state to protect the life, liberty, and property of the people.[21]

The 10 November 2016 opinion and order

On 10 November 2016, the Honorable Ann Aiken handed down her opinion and order: the plaintiffs had successfully alleged injury in fact, and the case would move to trial.[22] Specific harms to personal health and well-being resulted from climate-induced forest fires, droughts, extreme flooding, damage from hurricanes, and loss of snowpack.[23]

It was also determined that the plaintiffs satisfied causation. They were able to show a chain of causation between the greenhouse gas emissions produced in the United States, the power that the government has over increasing and decreasing those emissions, and how they use that power to engage in activities that promote increased fossil fuel combustion.[24] High emissions levels cause climate change, and climate change causes plaintiffs' injuries.[25]

The Fifth Amendment of the U.S. Constitution provides that 'No person shall be ... deprived of life, liberty, or property, without due process of law'.[26]

The plaintiffs have alleged that 'a stable climate system is a necessary condition to exercising other rights to life, liberty, and property'.[27] For the plaintiffs to successfully argue that a stable climate system is a fundamental right, they turned to *Obergefell*, 'which described the concept of liberty as evolving and not limited to the liberty interests that existed during the time of our founders'.[28]

Judge Aiken held that:

> In this opinion, this Court simply holds that where a complaint alleges governmental action is affirmatively and substantially damaging the climate system in a way that will cause human deaths, shorten human lifespans, result in widespread damage to property, threaten human food sources, and dramatically alter the planet's ecosystem, it states a claim for a due process violation. To hold otherwise would be to say that the Constitution affords no protection against a government's knowing decision to poison the air its citizens breathe or the water its citizens drink … Plaintiffs have adequately alleged infringement of a fundamental right to a stable climate system [which is] fundamental to a free and ordered society.[29]

Judge Aiken then outlined the Plaintiffs' public trust arguments, that 'The overarching public trust resource is our country's life-sustaining climate system, which encompasses our atmosphere'.[30] Therefore, 'As sovereign trustees, defendants have a duty to refrain from 'substantial impairment' of these essential natural resources'.[31]

She explained that Locke's social contract theory 'heavily influenced Thomas Jefferson and other Founding Fathers'.[32] As noted above, it 'provides that people possess certain inalienable rights and that governments were established by consent of the governed for the purpose of securing those rights. Accordingly, the Declaration of Independence and the Constitution did not create the rights to life, liberty, or the pursuit of happiness - the documents are, instead, vehicles for protecting and promoting those already-existing rights'.[33] And that 'the right of future generations to a 'balanced and healthful ecology' is so basic that it 'need not even be written in the Constitution for [it is] assumed to exist from the inception of humankind'.[34]

The opinion and order also discussed some of the broader issues of the failure of the state, through its environmental laws, to protect the environment, or in other words, the failure of the state to protect life. She noted that, 'Federal courts too often have been cautious and overly deferential in the arena of environmental law, and the world has suffered for it'.[35]

> The current state of affairs … reveals a wholesale failure of the legal system to protect humanity from the collapse of finite natural resources by the uncontrolled pursuit of short-term profits … [T]he modern judiciary has enfeebled itself to the point that law enforcement can rarely be accomplished by taking environmental predators to court.[36]

In May 2017, the federal magistrate recommended the denial of the Defendants' motion to dismiss, stating that 'the alleged valuing of short term economic interest despite the cost to human life, necessitates a need for the courts to evaluate the constitutional parameters of the action or inaction taken by the government. This is especially true when such harms have an alleged disparate impact on a discrete class of society'.[37] Trial begins 5 February 2018.

Global trends in state responsibility for climate change

> The costs of inaction are terrifying.[38]

There is momentum around the world for citizens to hold their governments accountable for climate change action or inaction. In *Urgenda Foundation v. The State of the Netherlands*, 900 plaintiffs across generations accused the Dutch government of 'knowingly contributing' to a breach of the 2°C maximum target for global warming.[39] Similar to *Juliana*, the legal arguments rested on duty of care, the duty not to harm (particularly forbidding states from damaging other states), and the precautionary principle.[40] On 24 June 2015, The Hague District Court ruled in favour of the Urgenda Foundation, finding that 'the severity and scope of the climate problem make it necessary for the Dutch government to reduce the Dutch greenhouse gas emissions at a much higher pace'.[41]

The Court also noted their role in global emissions and that 'there is … a sufficient causal link between the Dutch greenhouse gas emissions, global climate change, and the effects (now and in the future) on the Dutch living climate'.[42] The Oslo Principles on Global Climate Change Obligations were influential in the judge's reasoning.[43] The Dutch Government is appealing the case and the hearing at The Hague Court of Appeal is scheduled for 28 May 2018.

In November 2015, when a law student in New Zealand filed suit against her government in the High Court of Wellington for failing to do enough to meet their climate change targets, then-Prime Minister John Key dismissed the legal action as 'a joke'.[44] In *Thomson v. Minister for Climate Change Issues*, Sarah Thomson asserts that the Government failed to review and set targets that would 'stabilize greenhouse gas concentrations in the atmosphere at a level that would prevent dangerous anthropogenic interference with the climate system'.[45] In addition, the Government failed to consider 'those that are particularly vulnerable to the adverse effects of climate change'.[46] The judgment is pending in the High Court of New Zealand.

On 29 June 2016, seven-year-old Rabab Ali, on behalf of all the Pakistani people, filed a climate change lawsuit against the Federation of Pakistan and the Province of Sindh in the Supreme Court of Pakistan.[47] The Petition asserts that, 'through the exploitation and continued promotion of fossil fuels, in particular dirty coal, the Pakistan and Sindh governments have violated the Public Trust Doctrine and the youngest generation's fundamental constitutional rights to life, liberty, property, human dignity, information, and equal protection of the law'.[48]

Another outstanding case has recently come from India's National Green Tribunal, *Ridhima Pandey v. Union of India and the Central Pollution Control Board*.[49] On 22 March 2017, nine-year-old Ridhima Pandey filed a petition asserting that the Indian government had failed to fulfil its duties to her and the Indian people to mitigate climate change by failing to implement various environmental laws. Pandey also invoked the Public Trust Doctrine, the principle of sustainable development, the precautionary principle, and the inter-generational equity principle. The petition also highlighted the particular vulnerability of children to suffer the 'social, emotional, and cognitive impacts of climate change'.[50]

Rahul Choudhary, the attorney for Pandey, said:

> Ms. Pandey is a compassionate child who is fulfilling her duty as a citizen of India. She is simply asking her government to fulfill its own duty to protect the vital natural resources on which she and future generations depend on for survival.[51]

In mid-October, following Hurricane Ophelia, the strongest Eastern Atlantic hurricane ever recorded, hit Ireland. Motivated by global climate change litigation like *Juliana* and *Urgenda*, Friends of the Irish Environment filed suit against the Irish government:

> It's a case for everyone in Ireland, young and old. We're hoping the case will capture imaginations here and abroad and galvanize a movement pushing for ambitious action. The extreme impacts of climate change are beginning to hit home – we need to act urgently to ensure this is not the new normal for us and for our children and grandchildren.[52]

Friends of the Irish Environment v. Ireland obtained permission from the High Court on 24 October 2017 to proceed with the lawsuit.[53]

In June 2017, dry thunderstorms caused deadly wildfires across the Pedrógão Grande community in Portugal.[54] As a result, seven children aged 8 to 18 announced in October that they are organizing a lawsuit against 47 European nations of the Council of Europe for failing to take action on climate change.[55] They are asking these nations, as signatories to the Paris Agreement, to enforce stronger emission-cutting policies, and to stop mining fossil fuel reserves.[56]

Some lawyers and scholars are also seeing developments in international law that could help bolster these types of cases. For example, in September 2016, the International Criminal Court expanded their prosecutorial power to include environmental crimes.[57] In addition, the ICC Prosecutor launched the Policy on Children, calling it 'one of the key strategic goals of the Office'.[58] The Prosecutor holds that, 'A crime against a child is an offence against all of humanity; it is an affront to our basic tenets of human decency. Children are our greatest resource, and must be protected from harm so as to reach their full potential.'[59]

In addition, the UN Human Rights Council passed Resolution 32/33 'to prepare, in consultation with Member States and other relevant stakeholders, a detailed analytical study on the relationship between climate change and the full and effective enjoyment of the rights of the child'.[60] It has also been argued that the failure of states to act on climate change violates the UN Convention on the Rights of the Child, as climate change disproportionately threatens children.[61]

Conclusion

> I stepped right in the middle of climate change.[62]

With each wildfire, with each hurricane, with each flood, with each drought, the concrete impacts of climate change on children, families, businesses, and governments around the world are becoming more and more visible. But unfortunately, some states are not responding quickly or effectively enough. Unless a state is without the resources or infrastructure to act, most inaction is due to the pursuit of short term economic interests, personally or publicly, that ultimately put our life systems at risk.

Aristotle wrote that '[the city-state] comes into existence for the sake of life, it exists for the good life'.[63] This article has attempted to show that there are ethical and legal implications to state non-action when protecting the foundations of life, when the state effectively becomes the adversary to the environment. It is grounded in concepts of the primary duty of the state: to protect. From Hobbes's *Leviathan* to Locke's social contract theory; and for the United States, from the Declaration of Independence to the Federalist Papers, to the post-Civil War Reconstruction Amendments. The governance systems we structure, the laws we choose to create, implement, and enforce, or choose not to create, implement, or enforce, it is and always will be about our values, our ethics. Therefore, the citizens must always stand vigilant on where the values of their leaders take them, and their state.

Notes

1 John Locke (1698) 'Of the Ends of Political Society and Government', §123 and §131, respectively, chapter IX of *Two Treatises of Government*, http://oll.libertyfund.org/titles/locke-the-two-treatises-of-civil-government-hollis-ed.
2 Steven J. Hyman (1991) 'The First Duty of Government: Protection, Liberty and the Fourteenth Amendment', *Duke Law Journal*, vol 41, p. 515 and Locke at §131.
3 See generally Locke at chapter XIX §211–243; and quoting §211.
4 Barack Obama (28 August 2006) 'An Honest Government, A Hopeful Future', presented at the University of Nairobi, http://obamaspeeches.com/088-An-Honest-Government-A-Hopeful-Future-Obama-Speech.htm. This quote continues, 'And this is why the struggle against corruption is one of the great struggles of our time'. For further exploration into the links between corruption, governance, and the environment, see generally, Kathryn Gwiazdon (2015) 'Seeking Justice in a Land without Justice: The Application of Anti-corruption Principles in Environmental Law' in

L. Westra et al. (eds), *Ecological Systems Integrity: Governance, Law, and Human Rights*, Routledge, London.

5 Charles Sumner, United States Congress (18 January 1867) *The Congressional Globe: Containing the Debates and Proceedings of the Second Session of the Thirty-Ninth Congress*, p. 542 (regarding the Tenure of Office Act), www.digital.library.unt.edu/ark:/67531/metadc30916/. Note that this quote has been repeatedly mis-cited in publications as 'CONG. GLOBE, 39th Cong., 2d Sess. 101 (1867) (remarks of Rep. Farnsworth) (debating Reconstruction Act of 1867)'. The author discovered that the original mis-citation occurred in Hyman, Steven J. (1991) 'The First Duty of Government: Protection, Liberty and the Fourteenth Amendment', *Duke Law Journal*, vol 41, p. 508.

6 See generally www.senate.gov/artandhistory/history/common/briefing/Impeachment_Johnson.htm

7 Hamilton, Alexander or Madison, James (19 February 1788) 'No. 57 The Alleged Tendency of the New Plan to Elevate the Few at the Expense of the Many Considered in Connection with Representation' *The New York Packet*.

8 Thomas Hobbes (1881 reprint of 1651) *Leviathan; Or, The Matter, Forme, and Power of a Commonwealth*, p. 3, J. Thornton, Oxford

9 Pope Francis (24 May 2015) 'Encyclical Letter, Laudato Si' of the Holy Father Francis on Care for our Common Home', http://w2.vatican.va/content/francesco/en/encyclicals/documents/papa-francesco_20150524_enciclica-laudato-si.html.

10 Bruce Stokes et al. (13 June 2016) 'Europeans See ISIS, Climate Change as Most Serious Threats', *Pew Research Center, Global Attitudes and Trends*, www.pewglobal.org/2016/06/13/europeans-see-isis-climate-change-as-most-serious-threats/

11 Jennifer Hansler (8 November 2017) 'As Syria Joins Paris Climate Agreement, US Stands Alone', *CNN*, www.cnn.com/2017/11/07/politics/syria-paris-climate-agreement/index.html

12 Ann Aiken (10 November 2016) *Kelsey Cascadia Rose Juliana, et al., v. United States of America, et al.*, U.S. District Court, D. Oregon, Eugene Division, https://static1.squarespace.com/static/571d109b04426270152febe0/t/5824e85e6a49638292ddd1c9/1478813795912/Order+MTD.Aiken.pdf.

13 Supra note 12. The original case was filed on 12 August 2015 and amended 10 September 2015; https://static1.squarespace.com/static/571d109b04426270152febe0/t/57a35ac5ebbd1ac03847eece/1470323398409/YouthAmendedComplaintAgainstUS.pdf.

14 Julia Olson (30 August 2015) '21 Youths File Landmark Climate Lawsuit Against Federal Government', *EcoWatch*, www.ecowatch.com/21-youths-file-landmark-climate-lawsuit-against-federal-government-1882083365.html.

15 Supra note 12.

16 Thomas Coffin (28 June 2017) *Juliana*, https://static1.squarespace.com/static/571d109b04426270152febe0/t/59541c8db3db2b21ddf4c17e/1498684558660/2017.07.28+Order+Granting+Motions+to+Withdraw-Setting+Trial+Date.pdf.

17 Jayden Foytlin (7 September 2016) Declaration of Youth Plaintiff Jayden in Support of Plaintiffs' Opposition to Defendants' Motions to Dismiss, speaking on the hurricane floods that destroyed her home, https://static1.squarespace.com/static/571d109b04426270152febe0/t/57d0fac03e00be689aac4a09/1473313478990/JaydenDeclaration.pdf. Jayden's story highlights the courage and conviction of these Plaintiffs; she has been ostracized by a community that largely denies climate science, and sees it as a threat to the oil industry. Neela Banerjee and Zahra Hirji (13 June 2017) 'Fighting Climate Change Can Be a Lonely Battle in Oil Country, Especially for a Kid', *Inside Climate News*, https://insideclimatenews.org/news/13062017/kids-climate-change-lawsuit-childrens-trust-jayden-foytlin-louisiana.

18 See generally the Evolving Biosphere Ethic, http://environmentalethicsandlaw.org/the-evolving-biosphere-ethic.

19 Guest Contributor (30 August 2015) '21 Youths File Landmark Climate Lawsuit Against Federal Government', *EcoWatch*, www.ecowatch.com/21-youths-file-landmark-climate-lawsuit-against-federal-government-1882083365.html.
20 Id.
21 Our Children's Trust, 'Kids Name Trump as Defendant in Landmark Climate Case', *EcoWatch*, www.ecowatch.com/trump-kids-climate-lawsuit-2250707551.html.
22 Benjamin Hulac (30 June 2017) 'Kids' Climate Change Case to Go to Trial', *Scientific American*, www.scientificamerican.com/article/kids-climate-change-case-to-go-to-trial.
23 Supra note 12.
24 Supra note 12.
25 Supra note 12.
26 U.S. Const. amend. V.
27 Supra note 12.
28 Mia Hammersley (7 May 2017) citing *Obergefell v. Hodges*, 135 S. Ct. 2584, 2598 (2015) in 'The Right to a Healthy and Stable Climate: Fundamental or Unfounded?' *The Arizona Journal of Environmental Law and Policy*, www.ajelp.com/articles/the-right-to-a-healthy-and-stable-climate-fundamental-or-unfounded.
29 Supra note 12.
30 Id.
31 Id.
32 Id.
33 Supra note 12, citing Cf. Robinson Twp., 83 A.3d at 948.
34 Supra note 12.
35 Id.
36 Supra note 12, citing Alfred T. Goodwin, 'A Wake-Up Call/or Judges', 2015 Wis. L. Rev. 785, 785-86, 788 (2015).
37 Thomas Coffin (1 May 2017) *Juliana.*, US District Court, D. Oregon, Eugene Division, https://static1.squarespace.com/static/571d109b04426270152febe0/t/590793bb6b8f5bc7d71525da/1493668796765/17.05.01.Coffin+Order+Recommending+Denial+of+Interlocutory+Appeal.pdf.
38 Attorney Davey Salmon, *Thomson v. Minister for Climate Change Issues*, 'The Costs of Inaction are Terrifying', 27 June 2017, www.nzherald.co.nz/nz/news/article.cfm?c_id=1&objectid=11882663.
39 *Urgenda Foundation v. The State of the Netherlands* (Ministry of Infrastructure and the Environment), Hague District Court (24 June 2015), www.urgenda.nl/documents/VerdictDistrictCourt-UrgendavStaat-24.06.2015.pdf.
40 Arthur Nelson (25 June 2015) 'Dutch Government Ordered to Cut Carbon Emissions in Landmark Ruling', *The Guardian*, www.theguardian.com/environment/2015/jun/24/dutch-government-ordered-cut-carbon-emissions-landmark-ruling.
41 Supra note 41.
42 Id.
43 Oslo Principles on Global Climate Change Obligations, https://globaljustice.macmillan.yale.edu/sites/default/files/files/OsloPrinciples.pdf.
44 Ben Irwin (26 June 2016) 'Student Suing Government Says There Are "Logical Fallacies" in Climate Change Policy', www.newshub.co.nz/home/politics/2017/06/student-suing-government-says-there-are-logical-fallacies-in-climate-change-policy.html.
45 *Thomson v. Minister for Climate Change Issues*, High Court of New Zealand, Wellington, (10 November 2015), http://web.law.columbia.edu/sites/default/files/microsites/climate-change/files/Resources/Non-US-Climate-Change-Litigation-Chart/nz_case_statement_of_claim.pdf.
46 Id.

47 *Rabab Ali v. Federation of Pakistan*, Supreme Court of Pakistan (29 June 2016), www.elaw.org/system/files/Pakistan%20Climate%20Case-FINAL.pdf.
48 Naeem Sahoutara (29 June 2016) 'Seven-Year-Old Girl Takes on Federal, Sindh Governments', *Express Tribune*, https://tribune.com.pk/story/1133023/seven-year-old-girl-takes-federal-sindh-governments.
49 *Ridhima Pandey v. Union of India and the Central Pollution Control Board*, National Green Tribunal (22 March 2017), https://static1.squarespace.com/static/571d109b04426270152febe0/t/58dd45319f74568a83fd7977/1490896178123/13.03.22.ClimateChangePetition.pdf.
50 Id.
51 Our Children's Trust (30 March 2017) 'Youth Files Climate Case with India's Environmental Court: Youth-Led Climate Lawsuits Grow Internationally', https://static1.squarespace.com/static/571d109b04426270152febe0/t/58dd78f5f7e0abe149e9fb35/1490909429734/2017.03.30+India+Climate+Case+PR.pdf.
52 See www.friendsoftheirishenvironment.org/climate-case.
53 Id.
54 Julia Jones et al. (19 June 2017) 'Portugal Wildfire: 62 Killed, Victims Burned in Cars as They Fled', *CNN*, www.cnn.com/2017/06/18/europe/portugal-fire/index.html.
55 Ganguly, Manisha (19 October 2017) 'Children to Sue European Countries Over Climate Change', *CNN*, www.cnn.com/2017/10/19/europe/portugal-children-climate-change/index.html.
56 Id.
57 Office of the Prosecutor (15 September 2016) Policy Paper on Case Selection and Prioritisation, *International Criminal Court*, www.icc-cpi.int/itemsDocuments/20160915_OTP-Policy_Case-Selection_Eng.pdf.
58 ICC Press Release (18 November 2016), www.icc-cpi.int/Pages/item.aspx?name=pr1257.
59 Id.
60 See generally, www.ohchr.org/EN/Issues/HRAndClimateChange/Pages/RightsChild.aspx.
61 Convention on the Rights of the Child, www.ohchr.org/Documents/ProfessionalInterest/crc.pdf.
62 Supra note 17.
63 Aristotle Pol. 1.1252b, www.perseus.tufts.edu/hopper/text?doc=Perseus:abo:tlg,0086,035:1:1252b.

2 Critical and analytical considerations on climatic ethics

Marco Ettore Grasso

Introduction: ethical responsibility for climate change

Climate change puts us in the face of new theoretical categories concerning responsibility. In effect, the responsibility for this particular change is first of all a global responsibility, as it develops in space and time. It is not necessarily assumed on a causal link, since individual actions are not a necessary and sufficient cause of the global effect considered.

Nevertheless, the same responsibility is also contributory, because unsustainable human actions, such as driving a car, contribute anyway, albeit minimal, to the temperature rise. From this point of view, this responsibility is shared, namely a 'common responsibility'.

And last but not least, it is essentially a responsibility that takes into account the consequences of climate impacts and that we could therefore call 'consequential'. For example, we think of the effects that climate change determinates on human health.

Furthermore, on the subject of equitable emissions, several principles are relevant, such as the principle of the guaranteed minimum, the principle of the greater capacity, the principles of equal percentages, burdens and sacrifice, the principle of equal per capita parts and the common but differentiated responsibilities principle. Among these, the most important principle certainly is the common but differentiated responsibilities principle; hence, the analysis of the role of differentiation in the equitable emissions theories and the critical and analytical study of fundamental concepts in the climate change ethics.

Equitable emissions and differential treatment

In the *White Paper on the Ethical Dimensions of Climate Change*, Donald Brown (2013) presented several ethical approaches, the most relevant of which is inherent in the equitable allocations of per capita emissions and it is based on the idea that all human beings should be aimed at equitable sharing of the common good 'atmosphere'. The approach based on equity and applied to greenhouse gas emissions also requires that all individuals have the right to emit an equitable amount of carbon dioxide.

The common but differentiated responsibilities principle originally appeared in the Charter of Principles (specifically Principle No. 7) drawn up at the International Conference on Environment and Development held in Rio de Janeiro in 1992, which states that States have common but differentiated responsibilities because of different contributions to global environmental degradation. Article 3 of the Framework Convention on Climate Change, instead, identifies this differentiation in respective capabilities.

In my opinion, the principle of common but differentiated responsibility, under an analytical profile, sounds like a contradictory principle. The attributes that distinguish this responsibility, in effect, are 'common' and 'differentiated', whereby two antithetical adjectives with each other. The choice of these two attributes, however, is not random because it reflects the conceptual vagueness of the literature on climate change.

Nonetheless, 'equitable differentiation' constitutes one of the main conceptual categories useful to promote justice in environmental and climatic ethics. Differential treatment should be available only up to the extent that inequalities are homogeneously reduced. Otherwise, an inequity regime would be created and this treatment would lose meaning.

Moreover, from a substantial point of view, the political geography of territories has changed as a result of economic, political and financial crises – namely, on the one hand, countries that are becoming richer (such as China) and, on the other hand, increasingly indebted countries (such as Italy or Greece). Therefore, the vague and generic 'North–South' categorization is no longer valid for defining differentiation criteria.

The need of differentiation has been confirmed by the World Trade Organization (WTO), even through the Trade-Related Aspects of Intellectual Property Rights Agreement (1994), and by the Summit for the adoption of the post-2015 development agenda (2015). As early as the Bali Conference (2007) was directed to prepare a new agreement that provided for an operational model of international equity in line with the principle of common but differentiated responsibilities.

The combination between inequality and differentiation is underlined by Philippe Cullet (2016), according to whom the development of differentiation can be explained from two different perspectives. Firstly, differential treatment is founded on a recognition that severe inequalities must be addressed to ensure the legitimacy of the international legal order. Equity, in this sense, is the base of measures that seek to promote substantive equality in a world structured around formal equality. Secondly, differentiation is the outcome of the convergence of numerous interests manifested in international negotiations.

The conceptions of justice that underlie differentiation are mainly two: corrective justice and distributive justice. Corrective justice leads to focus the differential historical contributions of States to environmental degradation. For Posner and Sunstein (2007), climate change represents an authentic case of corrective justice: in effect, an evident correlation between emissions produced over the last few centuries and current levels of per capita economic development exists.

Distributive justice indeed addresses existing inequalities in human development and it suggests that is not enough to provide for equality of chances, but that what matters is equality of results (Fukuyama 2011).

Differentiation was also explained by the criterion of quality of emissions. In Shue's opinion a certain level of emissions is useful to achieve a minimum industrial development, suitable for combating poverty. In his work 'Subsistence Emissions and Luxury Emissions', the author has been asking what is the fair distribution of the costs needed to adapt to the social consequences of inevitable climate change (Shue 1993). In this regard, he distinguished between luxury emissions, necessary to secure luxury goods, and subsistence emissions, essential to provide useful goods for livelihood. Although it is not always easy to determine which goods are luxurious or of mere subsistence, ethically speaking the first type of emissions should have a significantly lower value than the second. In other words, according to Shue (ibid.) a criterion for the distribution of the costs of climate change that guarantees an equal development right would consist of different interdependent standards.

Posner and Weisbach (2010), for their part, believe that if the carbon dioxide emissions were combined with the impact of deforestation and urbanization, from 1950 to today, China, Russia and Indonesia would be at the top for historical contribution to global warming (after the United States). If we calculate the per capita emissions in the same period, we would find instead at the top countries such as Belize, Guyana and Luxemburg, characterized by inadequate production facilities.

The widely applied distribution criterion for greenhouse gas emissions is the so-called 'grandfathering', based on the historic emission levels of States (Kyoto Protocol, Article 3 (7)). In other words, this criterion states that equitable sharing of emissions for each party should be a function of its past sharing of emissions. The historical approach is also based on retributive justice, according to which mitigation efforts would be distributed among the polluting countries since they were made aware of the pollution phenomenon. In light of this, I think some opinions, including that of Bodansky (2004), are too much 'extremist'. The author thinks in fact that industrialized countries need to revisit their own share of emissions 'even more' than they would have been required by virtue of their historical responsibilities.

I generally believe that every generation will enjoy the benefits of the unjust actions of its own ancestors, but at the same time it is morally responsible for current pollution, regardless of the past.

I am considering below two concepts that occupy, in an absolutely impressive way, the ethical scene on climate change, that is to say the concepts of 'equity' and 'vulnerability'.

Equity: a theoretical premise

Some studies on equity exist in the scientific literature pertaining sustainability, especially with regard to intergenerational equity and equity understood as a means through which to share the burdens of global warming. Political and legal sources which consider equity as a fundamental matrix of sustainable development are numerous; for example, the Brundtland Report (1987), the New Delhi Declaration of Principles of International Law relating to Sustainable Development (2002), Art. 3 of the Framework Convention on Climate Change (1992), Principle 7 of the Charter of Principles (1992), the last IPCC Report (2014) and Art. 16 of the Desertification Convention (1994).

According to my view, the true meaning of equity should be identified in its original historical significance. In effect, 'equity' is a term derived from the Latin '*Aequitas*'. In the past, it was considered as a principle of divine inspiration. The glossers, ancient scholars of law, identified in God the sources of this particular 'justice', starting from a few moral ethical standards of religious matrix, they reworked the comprehensive system of rules of Roman law. Even under the common law system, equity regime plays a key role in the history of cases law.

Fairness framework, instead, concerns an impartiality of treatment and accordingly distribuitive justice issues are more pertinent, in line with the Rawlsian philosophy.

Currently, different interpretations of this concept emerge. For Judge Weeramantry, for instance, equity assumes the role of 'mitigation of the application of rigid rules'. He quotes Aristotle's *Nicomachean Ethics*, according to which the essential nature of what is right is the rectification of the law, where the law is deficient because of its general character. Equity in this case is seen as an ethical assessment of a non-moral but legal nature (International Court of Justice, Gabcíkovo-Nagymaros Project (Hungary/Slovakia), 1997, par. 118 (d)).

In another decision, instead, the International Court of Justice ruled that equity as a legal concept is 'a direct emanation of the idea of justice' (International Court of Justice, Continental Shelf (Tunisia/Libyan Arab Jamahiriya), 1982, par. 71).

Equity regime in the climate change literature

In the field of climate change, this expression can be understood in terms of distribution, that means as that situation in which 'every person has a right to the same level of greenhouse gas emissions as every other person' (Jamieson 2005).

Originally, equity issue focused largely on the concept of 'mitigation': first works (e.g. Shue 1993) confined equity in terms of greenhouse gas emissions production and partition of the burden of emission reductions; hence, arguments on 'distributive justice', 'causal' and 'historical' responsibility, 'vulnerability', 'need' and 'ability' arose in this field. The study of the differences between Parties' contributions to emissions and abilities to reduce the same emissions was heading toward the adoption of the common but differentiated responsibilities principle, a principle that is dominant in the context of the international operationalization of equity. In recent times, instead, most of the proposals for a new post-2012 agreement focused on 'adaptation' framework (Adger et al. 2006).

Often the term equity is used interchangeably with other terms such as 'justice' or 'fairness. Only by way of example, I mention Pickering et al. (2012), who use synonymously the terms 'fairness' and 'equity' as broader moral and political concepts applicable to negotiations. They believe in fact that both fairness and equity may apply to the substantive distribution of goods as well as procedures for distribution.

According to IPCC (2014), 'equity' is an integral dimension of sustainable development and it is supposed by three criteria:

1. a 'moral' justification that draws upon ethical principles;
2. a 'legal' justification that appeals to existing treaty commitments and soft law agreements in order to cooperate in accordance with equity principles;
3. an 'effectiveness' justification by virtue of which a fair arrangement is more likely to be agreed internationally and successfully implemented domestically.

However, the question of the role that equity should play in the establishment of global climate policy and burden sharing in particular is certainly controversial. Nevertheless, IPCC (2014) identifies some principles as corollaries of equity. These principles are: 'responsibility', 'capacity', 'right to development' and 'equality'.

IPCC and equity: conceptual confusions

Both tipping points and advocates of a 'equity regime' in the climate negotiations recognize that governments can choose to act on moral rather than purely self-interested principles (Baer 2013).

The IPCC scholars propose two theoretical typologies of 'equity': (a) 'distributive equity' and (b) 'procedural equity'. Distributive equity should concern a distribution of resources in contexts such as burden sharing, as well as a distribution of well-being in the broader context of social justice. Procedural equity, instead, is linked to the framework of participation in decision making.

Also IPCC (2014) admits the conceptual confusion concerning 'equity', 'justice' and 'fairness'. In chapter 4 of the report, in fact, IPCC hightlights: 'in this chapter the terms equity, fairness and justice are not distinguished, but are used according to common usage depending on context' (ibid.).

Chapter 3 of the report, then, dedicates to this particular 'tipping point' the 'FAQ' number 3.2, entitled: 'Do the terms justice, fairness and equity mean the same thing?'. In this chapter it is said that the terms 'justice', 'fairness' and 'equity' are used 'with subtly different meanings in different disciplines and by different authors', without however specify what disciplines and authors (ibid.). In addition, the text specifies that 'justice' is used more frequently in philosophy, while 'equity' in social science. Even in this case, there is no demonstration, through scientific quotations, that equity is a theoretical category predominantly of social sciences rather than ethics. This finding appears, once again, completely superficial, without foundation. Furthermore, equity is an issue primarily related to the philosophical-moral aspect that affects the distributive element of greenhouse gas emissions and therefore related to an ethical category.

The conceptual confusion noted by the IPCC culminates under the same chapter in the finding according to which 'in reporting on the literature, the IPCC assessment does not impose a strictly uniform usage on these terms. All three are often used synonymously' (ibid.). This affirmation confirms the scientific superficiality with which IPCC approaches the equity theme. Recognizing confusion on the conceptual level, this scientific body does not address the problem, but instead wants to remain consciously in that confusion.

Notwithstanding, it tries to give a definition of 'fairness' and 'justice', whereas justice is defined in their chapter 3 as 'broadly concerned with a person receiving their due' (ibid.). Even in this case, the approximation and the difficulty of expressing a scientific definition seems quite obvious. With respect to fairness, indeed, the report specifies that sometimes it is used in the narrower sense of receiving one's due (or 'fair share') in comparison with what others receive. The vagueness and incomprehensibility of this definition is fortunately attenuated by an example, which is concluded with the statement that fairness is concerned with the distribution of goods and harms among people.

Complicating this approximate picture, we find paragraph 3.3.3 of chapter 3, which speaks of distributive justice. And without deepening the theory of justice closer to climate change matter, it affirms that among the widely discussed views on distributive justice there are strict egalitarianism and indirect egalitarian views, including prioritarianism and sufficientarianism. With this definition, the conceptual framework is greatly complicated, because the theories invoked by the IPCC are largely those relating to the principle of 'equality'. Thus, apart from the declared conceptual confusion between equity, justice and fairness, a new term that could be used interchangeably with the previous words is added, that is the concept of 'equality'.

End-point and starting-point interpretations of the concept of vulnerability in social science of disasters

According to the IPCC Fifth Assessment Report (2014), vulnerability is defined as the propensity or predisposition to be adversely affected. Vulnerability encompasses a variety of concepts and elements, including, although in a broad sense, sensitivity or susceptibility to harm and lack of capacity to cope and adapt. Therefore, in accordance with IPCC (chapter 19), a set of factors (wealth, social status and gender) determines vulnerability and exposure to climate-related risk.

In the same chapter, IPCC scholars say that vulnerability refers primarily to characteristics of human or social-ecological systems exposed to climatic hazards, such as droughts and floods or non-climatic events and trends (increasing temperature and sea level rise). Despite this, I think it is significant that 'health' is not considered among the factors that determine vulnerability. In fact, health can lead to vulnerability, considering for example that climate change greatly affects health. A compromised health, even from damage caused by pollution or greenhouse gases, can certainly be a factor that causes a greater vulnerability.

The IPCC Third Assessment Report that was published in 2001 instead provided a more simplified definition of 'vulnerability', considered as a function of exposure, sensitivity and adaptive capacity.

The future trend is that of associating increasingly the concept of vulnerability to the science of disasters and catastrophes (Birkmann, Sorg and Welle 2017; Grasso 2017).

From an analytical point of view, the vulnerability concept can be examined in accordance with two different interpretations of meaning. In line with a first interpretation, vulnerability is seen as the future consequence of climate impacts, where current adaptive capacity goes towards the subsequent vulnerability, understood as end-point. Within this dimension, the abatement of greenhouse gases emissions, as well as the attenuation of the sensitivity that climate change impacts determinate on several systems are the primary problem resolution. This theoretical interpretation of the concept of vulnerability has been identified as 'end-point' (O'Brien et al. 2004).

According to some authors (Brooks et al. 2005), this perspective does not consider socio-economic aspects, focusing mainly on prior conditions and not on future stresses. For this purpose, Kelly and Adger (2000) recall the image of the 'wounded soldier': this metaphor assumes that the vulnerability of individuals or communities to climate hazards is caused by their 'capacity to respond to that hazard, rather than by what may or may not happen in the future'.

This critical point of view is directed to the enhancement of another interpretive mode of vulnerability, which sees vulnerability as a 'starting-point'. Within this dimension, vulnerability motivates and determinates adaptive capacity.

There is, however, a 'tertium genus', which is a third possibility of interpreting vulnerability. In my opinion, in effect, the two different interpretations of the concept of vulnerability should not be separated from one another. They complement each other. On the one hand, it is true that the impacts caused by climate change generate a future vulnerability, however it is also true that vulnerability, caused not only by climate impacts but also by socio-economics preexisting factors, exacerbates due to an inadequate ability to deal with extreme events.

Besides, I think the vulnerability concept rotates around two corollaries, which do not relate so much to the 'time status' of the vulnerability compared to climatic impacts, resulting in an end-point or a starting-point vulnerable status, but rather they concern the human, territorial and social condition of the community concerned. This condition can be 'fragile' or linked to a 'skill' that is capable of adapting and preparing for future climate impacts. The first corollary therefore adheres to the 'fragility' of a community and a territory, the second one instead concerns the ability to react, namely the 'ability' of a population to create adaptive capacities. Therefore, I would like to call these two theoretical groups with the expressions 'fragility vulnerable status' and 'ability vulnerable status'.

Within the 'fragility vulnerable status', terms such as 'sensitivity', 'propensity', 'probability of occurrence', 'predisposition' to be affected by climate impacts, 'exposure' to shocks, alarms and stresses are consistent with this status. In this sense, sensitivity sees vulnerability in terms of the 'probability of occurrence', as negative impacts of weather and climate-related events.

In the 'ability vulnerable status', instead, the role of adaptive capacity is essential. This concept, which is different from that of 'adaptation', is linked with other concepts, such as those of 'resilience' and 'ability to cope with' climate impacts.

Conclusions

Ethical responsibility for climate change is a moral and political issue; however, it is not always easy to identify. Climate change literature, also in its interdisciplinary nature, shows a conceptual framework which, under an ethical-analytical perspective, is rather incomplete.

Rarely IPCC is contested, because it is considered the most authoritative body in this field. Nevertheless, I think that language is very important for every science and therefore even for the climate change science. The declared superficiality by IPCC is not permissible. Many of its definitions appear vague and superficial.

In climate change literature, the concepts of 'equity' and 'vulnerability' are subject to several interpretations and this generates a lot of confusion. They are core terms of climatic ethics. Even the concept of 'climatic refugees' deriving from the notion of 'environmental refugees' is quite controversial due to numerous definitions of the same notion, so to create a sort of conceptual confusion. According to the Organisation for Economic Co-operation and Development, for example, these refugees are people forced to emigrate for environmental reasons, degradation, disappearance of lands or natural disasters. The International Organization for Migration, on the other hand, defines them as persons who, because of sudden or gradual environmental changes adversely affecting their living conditions, are forced to abandon their homes or choose to do so temporarily or permanently, moving within the respective country or crossing the national borders. In accordance with a narrower view, instead, the International Institute for the Environment and Development compares these refugees to seasonal migrants.

Moreover, within a moral and substantial perspective, the scientific works should more investigate the 'limits' of climate adaptation. In my opinion, among these limits we can find the 'gap' between 'time' variable ('slowness') which addresses the development of national adaptation measures and the velocity of the effects produced by extreme events ('rapidity'). We are witnessing, in effect, more and more destructive extreme events, where the prediction of the risks of these same events is not always so easy to practice (sometimes, indeed, the worst disasters caused by extreme events are those compared to which the risk of verification is quite low).

Finally, adaptation funds are not always seriously based on a specific vulnerability assessment of the country concerned. An ethical relationship between social complexity, vulnerability and capacity should be better investigated.

References

Adger, W. N., Paavola, J., Huq, S. and Mace, M. J. (2006) *Fairness in Adaptation to Climate Change*, MIT Press, Cambridge, MA.

Birkmann, J., Sorg, L. and Welle, T. (2017) 'Disaster Vulnerability', in M. Pompella and N. Scordis (eds), *The Palgrave Handbook of Unconventional Risk Transfer*, Palgrave Macmillan, Basingstoke.

Bodansky, D. (2004) *International Climate Efforts Beyond 2012: A Survey of Approaches*, Pew Center on Global Climate Change, Arlington, VA.

Brooks, N., Adger, W. N. and Kelly, P. M. (2005) 'The Determinants of Vulnerability and Adaptive Capacity at the National Level and the Implications for Adaptation', *Global Environmental Change*, vol. 15, no. 2.

Brown, D. A. (2013) *White Paper on the Ethical Dimensions of Climate Change*, Legal Studies Research Paper no. 13-58, Widener University Delaware Law School, Wilmington, DE.

Cullet, P. (2016) 'Differential Treatment in Environmental Law: Addressing Critiques and Conceptualizing the Next Steps', *Transnational Environmental Law*, vol. 5, no. 2.

Fukuyama, F. (2011) *The Origins of Political Order: From Prehuman Times to the French Revolution*, Profile Books, London.

Grasso, M. E. (2017) 'Natural Catastrophes and Forms of Catastrophism: A New Ethical and Moral Framework Leading towards the Responsible Catastrophism Model', in L. Westra, J. Gray and F. T. Gottwald (eds), *The Role of Integrity in the Governance of the Commons*, Springer, Berlin.

IPCC (2014) *Climate Change 2014: Mitigation of Climate Change*, Cambridge University Press, Cambridge.

Jamieson, D. (2005) 'Adaptation, Mitigation and Justice', in W. Sinnott-Armstrong and R. B. Howarth (eds), *Perspectives on Climate Change: Science, Economics, Politics, Ethics*, Elsevier, Amsterdam.

Kelly, P. M and Adger, W. N. (2000) 'Theory and Practice in Assessing Vulnerability to Climate Change and Facilitating Adaptation', *Climatic Change*, vol. 47, no. 4.

O'Brien, K., Eriksen, S., Schjolden, A. and Nygaard, L. (2004) *What's in a Word? Conflicting Interpretations of Vulnerability in Climate Change Research*, Working Paper no 4, CICERO, Oslo.

Pickering, J., Vanderheiden, S. and Miller, S. (2012) '"If Equity's In, We're Out": Scope for Fairness in the Next Global Climate Agreement', *Ethics and International Affairs*, vol. 26, no. 4.

Posner, E. A. and Sunstein, C. R. (2007) 'Climate Change Justice', *Georgetown Law Journal*, vol. 96, no. 5.

Posner, E. A. and Weisbach, D. (2010) *Climate Change Justice*, Princeton University Press, Princeton, NJ.

Shue, H. (1993) 'Subsistence Emissions and Luxury Emissions', *Law and Policy*, vol. 15, no. 1.

3 Addressing climate change in a digital age

Rose A. Dyson

Introduction

New digital technologies underscore the extent to which we live in a global village. Good global governance grows increasingly urgent as issues pervade national boundaries. Public health concerns merge with those in education, the environment, energy, culture, gender, security and the economy. New media offer enormous potential in mobilizing for political action, limiting climate change, maintaining the biosphere and securing energy needs. But they can also be mobilized for harmful, destructive purposes.

We need to rethink the profit driven ways in which proliferating digital technologies mitigate steps toward a sustainable future. A more integrative approach to all policy making is essential to offset tensions arising from the social instability that rapid change is precipitating. This includes acknowledging and addressing growing health and public safety problems due to the harmful effects of media which encourage the use of violence as a conflict resolution strategy and reinforce inclinations toward hatred, racism, misogyny and fear. Moreover, in a myriad of ways that go well beyond the seductive and addicting trends of endless hours spent online, proliferating digital devices are neither carbon neutral, nor as some proponents would have us believe, do they hold the sole promise of innovation and job creation.

New technologies not neutral

Technological products are never neutral. Their invention and adoption by humans condition lifestyles, influence value systems and shape social possibilities, usually in the interests of the most powerful in society. As Marshall McLuhan pointed out, in our electronic age, a total new environment has been created (McLuhan 1964). Social instability brought on by rapid change, coupled with growing public awareness of the need to address climate change, underscores the urgency of re-examining ways in which we organize ourselves politically, economically, culturally and spiritually. But, ecologically sustainable culture cannot be reduced to partial responses to the immediate problems of pollution, environmental decay and natural resource depletion (Pope Francis 2015). We need an educational approach to generate resistance to the assault of the technocratic paradigm that dictates the profit driven terms by which we learn, communicate and amuse ourselves.

Government public policy must meet the test of its impact on all converging challenges such as automation, rapid change and climate change. It must also be screened through the lens of its social impact. Otherwise, new initiatives launched following the 2015 Paris Summit will become mired in the same market driven, globalized logic that equates economic health and prosperity with over consumption, trends toward economic inequality and disregard for degradation of our natural environment. These counterproductive tendencies are prevalent but largely overlooked as they apply to digital technologies.

Examining the medium for the message

All technology envelops us as we live out our lives in reconstructed, human-created environments. These developments accelerate our alienation from nature and, in turn, the destruction of nature by moving us farther inside an already pervasive artificial reality. They discourage self-sufficiency while reinforcing fear and disorientation. A tendency to grasp at simplified solutions is leading to a thirst for strong, autocratic leadership.

We have not yet grasped the fact that many technologies determine their own use, their own effects and even the kind of people who control them. McLuhan urged us to think of technology as having ideology built into its very form. In the early 1950s television was introduced to shift a wartime economy to one based on consumerism. At the heart of it all, was and still is the advertising industry, one of the most ubiquitous and tenacious forms of communication and ideology in society (Dyer 1982).

For decades, it has supported and largely determined, what we see, hear and read. More recently mainstream newspapers and television have experienced declining advertising revenue as it migrates to social media, compounding existing problems of biases in news coverage. Advertising influences the policies of all media, advancing and perpetuating the ideas and values indispensable to 'developed' economies. Its purpose is to persuade us to buy things, use them, throw them away and buy replacements in a cycle of continuous and conspicuous consumption. Clearly, such underpinnings are at odds with initiatives to encourage reducing, reusing and recycling goods to ensure a sustainable future. As Vance Packard pointed out over 50 years ago, in his provocative best sellers on the American philosophy and practices of deliberate waste under the pretence of making America great, such economic strategies drain our finances, undermine our children and threaten our future.

The silent crumbling of democratic institutions

Benjamin Barber says the triumph of capitalism has created a cultural ethos of induced childishness (Barber 2007). This infantilization is closely tied to the demands of consumer capitalism in a global market economy. Like Neil Postman (1986), Chris Hedges has also argued, that we must wake up before it is too late (Hedges 2009). We are now immersed in a cultural environment that has passively given up the linguistic and intellectual tools to cope with complexity and to separate illusion from reality (ibid.). Addicted to our digital toys, we are deluded into a lifestyle of endless texting inside social mediated worlds where virtual contact replaces real human contact. Online communities where like-minded individuals exchange messages in isolated silos encourage alienation from governing establishments and a polarization of opinions. They discourage a healthy exchange of ideas and can give rise to unrealistic hopes for political change. The 2016 Presidential election was a manifestation of these trends.

Falsehoods which permeated American presidential election propaganda in 2016 helped to perpetuate the notion that the truth is irrelevant. A sense of entitlement arose to 'one's own set of facts'. Facebook and social media assumed unprecedented influence as populist sloganeering, rife with disregard for factual data, pervaded cyberspace. Mainstream news media were relegated to secondary importance. But, the rise of populism is not entirely an outgrowth of ultra-nationalism, fascism and white supremacy. It is also a reaction to the failure of neoliberalism. For too long the system has served dominant corporate interests shrouded in the trappings of liberal democracy giving rise to anxieties and fears of being left behind, especially among the marginally educated, in a world where change is accelerating and challenges are multiplying.

Our culture is now saturated with images and slogans affording the mirage of functional literacy among youth who often choose not to read at all. Critical thinking skills are being replaced with superficial, technological operating skills. Propaganda masks ideas and ideology in the production of real knowledge. The predominant emphasis is on how we can be encouraged to feel good rather than to think critically. Most media scholars have concluded that if we understood the process better, we would learn to resist the seductive influences involved and have advocated age appropriate courses in media literacy throughout the educational process (Duncan et al. 1996).

Clearly that has not worked very well. Instead, the bar gets lower every year with new manifestations of media excesses dominating the preoccupations of the chattering classes, whether these revolve around how to handle reality TV billionaire celebrity aspirants for political office such as Donald Trump, corporate privileges of digital gurus who argue against any government regulation or public airwaves dominated by Hollywood trivia. Exploitive media industries have been given carte blanche to conduct business according to what feeds the bottom line, regardless of consequences. Their practices have been skilfully wrapped in rhetoric that ensures the public is conditioned to accept the premise that corporations are like individuals; entitled to freedom of expression regardless of its nature, be it political, pornographic or the seduction of children from the time they first come out of the womb.

But things are changing. Civics courses in American high schools are making enormous gains in popularity in the aftermath of the Trump election. *New York Times* journalist Timothy Egan points out that we may be witnessing a new, accidental renaissance in the United States with an awakening to the values under siege by a dark sided presidency. Subscriptions for mainstream newspapers have grown. The public has learned to appreciate the role of the judiciary in the American democratic process. Grass roots political activism is up and even spirituality (Egan 2017).

Commercial exploitation of children

'Pester power' and 'the nag factor' are techniques used by advertising agencies to target children with commodities that go far beyond the digital seduction which starts with 'brainy baby and Baby Einstein' videos, regardless of extensively documented research showing harmful effects. According to the Harvard Medical School-based, Campaign for a Commercial-Free Childhood (CCFC)in Boston, in 2006, Disney sold 20 million Baby Einstein videos despite the fact that both the American and Canadian Academies of Pediatrics recommend no screen time at all for children before the age of 2 years.

Over $17 billion was spent in 2006 by the advertising industry marketing products to children, an increase over $100 million spent in 1983. Every year, the average child 2–11sees 20,000 ads on television, not including product placement and over $500 billion is spent in purchases influenced by children under the age of 12 (see c-cave.org and commercialfreechildhood.org).

Most developed economies have introduced cooperative restraints on the harmful effects of advertising to minors. Advertising to children 13 years and under is banned in Quebec. Great Britain and most European countries have adopted similar legislation in response to concerns. These include the harmful effects of screen-based violence, sexual explicitness, advertising of junk food causing physical health problems such as obesity and heart disease, and cancer causing cell phones. It has yet to be passed in the rest of Canada, despite repeated calls from numerous Boards of Health, bills introduced by NDP members in Ontario and the House of Commons in Ottawa, and a call in a *Globe and Mail* editorial in 2016. Subsequently, Bill S-228, an Act to amend the Food and Drug Act which would prohibit food and beverage marketing to children 13 years has been introduced federally (Picard 2017).

Misplaced emphasis on economic growth and job creation

On climate change, many have pointed out, that our problem is not about carbon, but capitalism (Klein 2014). Unemployment, which is fuelling anxiety, disenchantment, terrorism and social instability, is, in part, brought on by digital advances geared predominantly toward feeding the bottom line (Friedman 2016). If we choose to we can seize the moment to transform our failed system and build something better. But this cannot happen without examining digital technologies which comprise a growing sector of the current economy. We could start by regulating the gaming industry. In 2010 *The Economist* predicted that video games would be the fastest growing form of mass media over the coming decade, estimated at around US$82 billion by 2015 (Economist 2010). The action-packed (code for violence filled) video game *Call of Duty: Black Ops* had fans in countries around the world queued for blocks on the first day of its release in 2010 – an example of what Benjamin Barber calls infantilized consumerism.

In 2011, it was reported in *The New York Times* that wasteful government spending in both Canada and the US involves generous tax breaks for video game producers regardless of content. Production and distribution of gory video games is now one of the most highly subsidized businesses on the continent. Government policy statements and annual budgets usually call for improving incentives for film and TV productions. But the nature of proposed production also needs to be examined. In 2008, a bill was brought before the House for the elimination of such incentives for audio visual productions involving extremely violent and pornographic content deemed to be harmful to the public interest. It was ultimately defeated in the Liberal dominated Senate. Industry lobbyists successfully advanced the argument that introduction of such discretionary funding would be too grave a threat to freedom of expression. The reality is that what it would impede is freedom of corporate enterprise regardless of the consequences. In December, 2016, the federal health committee was mandated to examine the public health effects of access and viewing of online violent and degrading sexually explicit material on children, women and men. It was the first examination of the problem since 1985.

So far, in Canada, any attempts to eliminate tax incentives for harmful productions have been futile. In 2015, it was reported in *The Globe and Mail* that of the $1.2 billion collected in carbon taxes in the province of British Columbia in 2014, 300 million was given to the audiovisual industries, regardless of content. Barry McKenna (2015) reported on how, back in 2009, the Ontario Government 'was ready to do almost anything to lure the burgeoning video game industry to the province'. The objective was to build an economy by spawning a new kind of manufacturing for job creation. As a result, Ubisoft S.A., the French maker of popular video games, such as *Assassin's Creed*, with imagery that mirror's the grizzly beheadings of American journalists posted online by IS in recent years, received the single largest business grant made by the province in the past decade.

Enthusiasm for the industry grows regardless of countless studies about the harmful effects of violence as entertainment done in the past, some of them by the Ontario Government itself (Dyson 1995, 2000). The Entertainm ent Software Association of Canada claims to generate $2.3 million annually in GDP. Said CEO Jamie Hitchie, 'I don't think a lot of people are aware of how prolific our video game industry is in the global context. We have more video game developers per capita than anywhere else in the world and we're producing some of the world's best titles.' Among them, he enthused, was *Assassin's Creed*.

Violent gaming fuels terrorism

Once one is on the path to violent extremism there is no end to what would be terrorists can find online to solidify their beliefs. According to Catherine Solyom, the Islamic State 'has made it simple for sympathizers worldwide to take up the extremist's cause' (Solyom 2015: A13). Conflict tourism is now prevalent among violent entertainment addicted gamers. In her article 'The road from rapper to ISIL pitchman', journalist Sarah Kaplan quotes one recruiter with a background as a Westerner and a musician in a posted video beckoning viewers to 'join the fun', as he rapped over grisly footage of Jordanian pilot Moath al-Kasasbeh being burning alive (Kaplan 2015: A13). Meanwhile, with each successive school lock down or terrorist attack precipitated by outbursts of violence, hand wringing about potential solutions and strategies for prevention among educators, law enforcement personnel and counter terrorism experts persists with scant consideration given to the well documented harmful effects of a steady diet of violent entertainment (Little 2016: A16; Bourrie 2016).

Promotion of *World of Warcraft* gaming, estimated to have well over10 million members paying US$15 a month, is bad for both the cultural and natural environment. Such gaming is designed to be addictive, considered the main indicator of success in the industry (Bakan 2011: 23). It is no accident that we have rising levels of youth gang violence, internet addictions, cyberbullying and evidence of mental instability in young adults. We have been warned for decades about how media violence contributes to what American psychologist, researcher and military expert David Grossman (2004) calls acquired violence immune deficiency syndrome (AVIDS). George Gerbner created an extensive body of literature and research measuring techniques that show how media violence leads to a mean world syndrome (Dyson 1995, 2000; Morgan 2002).

In October 2014 Anna Marie Tremonte hosted a programme on CBC Radio, *The Current*, addressing the harassment of women in the video game design field. No mention was made about the nature of the content and how it might actually pollute the work environment. The CEO spoke of how she enjoys 'action-filled' video games such as *Halo* and *Call of Duty*, and said a key strategy was for 'greater market penetration into the Middle East'. Evidence that IS is already extremely media savvy and uses social media to recruit and radicalize both male and female youth around the world was ignored. Such wilful blindness is prevalent throughout the mainstream media.

In recent decades, the discourse on morals and values education has shifted to moral relativism and postmodernism. Peter Knight at the School of Criminology at Simon Fraser University in British Columbia argues that the Trump Administration has adopted postmodernist thinking by signalling intensions to counter the mainstream media's version of truth with their own alternative facts (Knight 2017). According to postmodernist belief systems, what counts as truth in society, depends on one's own political or other interests. It is the modern university, the bastion of left wing thought, that has given birth to postmodernism and the idea that our ability to discover universal truth has been a failure. Conventional thinking is that we are all trapped within a perspective informed by our own beliefs, goals and aspirations. Until we get beyond such loose and personally selective notions of the truth and gravitate back toward a universal set of values and morals, the difference between facts and lies will remain debatable. Ultimately, objective facts with rules and regulations that we collectively agree upon form the cornerstone to advanced democratic governance.

Cancerous economic growth threatens health and well-being

The harmful effects of over exposure to digital technologies goes well beyond their impact on children. An enormous body of scientific evidence has accumulated about their addictive tendencies, sedentary nature of their use, tendency to impede development of socializing skills, apart from those cultivated online, contribution to attention deficit disorders and so on (Dyson 1995, 2000; Morgan 2002). Added to these have been warnings issued by the World Health Organization (WHO), former president of Microsoft Canada Frank Clegg and American scientist Devra Davis, about dangers from exposure to low level radiation from wireless devices and how these can increase the risk of cancer (Davis 2010). Since 2011, many countries which include Belgium, France, and India have passed laws either prohibiting the placement of cellular antennae on the roofs of hospitals, schools and playgrounds, restricting the sale of kiddie-phones designed for children or recommending limited exposure due to health risks. Cordless phones, Wi-Fi, smart meters and cell towers effectively function as low-level, constantly-emitting microwave transmitters. In Canada, the WHO warnings continue to be ignored by various levels of government.

Energy waste and insatiable demands

According to a report released by Greenpeace in 2014, digital technology products and services accounted for about 2 per cent of worldwide emissions in 2012, roughly the same as the airline industry. Some of the biggest electricity demands come from huge data centres that house the stacks of computers that process search requests, store photos and stream video. These online services dubbed 'cloud computing' – identified as the fastest growing sector of the digital industries – collectively consume more electricity than all but five countries – China, the US, Japan, India and Russia. While strides have been made by digital giants focusing on renewables, the emphasis on marketing practices that create demand continues unimpeded. Each advance depends on a further cradle to grave cycle of extraction and transport from manufacture to market – then from consumption to waste pit. These electronic cast-offs contain a witch's brew of heavy metals and toxic substances such as lead, cadmium and mercury. Some are recycled but most of are shipped off to developing countries willing to pay cash for trash.

Finally

The rise and continued emphasis on hyper consumerism characterized by inequalities, immaturity, indulgence, and childishness focused on faux needs impedes progress toward a sustainable future. The efforts of marketers to manipulate the needs and wants of gullible consumers cannot continue to be ignored either by climate action or social justice advocates. Like the nurturing of violence as a conflict resolution strategy, encouraging such infantilized behaviour is counterproductive. When he launched the Cultural Environment Movement at its founding convention in St Louis, Missouri, in 1996, with the help of over 150 independent organizations and supporters from 64 countries around the world (including this author), George Gerbner stressed that cultural activists needed to take lessons from the natural environmental movement. But what is needed now is not only for the current focus on climate change to be broadened to embrace the cultural environment movement but on human cognition as well. Rapid change, along with the urgent need for transition to a new, sustainable economy, must include attention to human fear and anxiety.

Media scholars, educators, school administrators and health professions must more vigorously and courageously resist the unimpeded explosion of digital devices that reinforce consumerism in some of its ugliest and most destructive forms. A more discriminating approach to every new invention and release on the market is essential. Such emphasis would help to move forward calls for a new economic order and an educational approach to address the technocratic paradigm that has led us to where we are today. Unimpeded economic growth, even with divestment from fossil fuel companies is only half the answer (Rubin 2015). Reining in a voracious industrial appetite for any kind of energy, clean or otherwise, for the purpose of producing and distributing goods no one needs and are often harmful in an age dominated by digital technology, is essential and long overdue.

References

Baken, J. (2011) *Childhood Under Siege: How Big Business Targets Children*. Allen Lane, London.

Barber, B. (2007) *Consumed: How Markets Corrupt Children, Infantilize Adults and Swallow Citizens Whole*. Norton, New York.

2016) *The Killing Game: Martyrdom, Murder and the Lure of ISIS*. HarperCollins, Toronto.

Davis, D. (2010) *Disconnect: The Truth about Cell Phone Radiation, What the Industry Has Done to Hide It, and How to Protect Your Family*. Dutton, New York.

Duncan, B., J. D'ippolito, C. Macpherson and C. Wilson. (1996) *Mass Media and Popular Culture*. Harcourt, Toronto.

Dyer, G. (1982). *Advertising as Communication*. Routledge, New York.

Dyson, R. A. (1995) The Treatment of Media Violence in Canada Since Publication of the LaMarsh Commission Report in 1977. *Doctoral thesis*, OISEUT, Toronto.

Dyson, R. A. (2000) *Mind Abuse: Media Violence in an Information Age*. Black Rose Books/ UT Press, Montreal.

Economist (2011) Special Report: Video Games. *The Economist* (10 December).

Egan, T. (2017) The New Accidental Renaissance. *The New York Times* (11–12 March): A14.

Friedman, T. L. (2016) *Thank You for Being Late*. Farrar, Straus and Giroux, New York.

Grossman, D. (2004) *On Combat: The Psychology and Physiology of Deadly Conflict in War and Peace*. PPTC Research Publications, New York.

Hedges, C. (2009) *The End of Literacy and the Triumph of Spectacle*. Alfred A. Knopf, Toronto.

Kaplan, S. (2015) The road from rapper to ISIL pitchman. *The National Post* (31 October): A13.

Klein, N. (2014) *This Changes Everything*. Alfred A. Knopf, Toronto.

Knight, P. (2017) Trump as Postmodernist: Truth No Longer Bound by Facts. *The Globe and Mail* (28 January): PF7

Little, C. (2016). How Can We Prepare for Violence if the Attacker is a Student? *The Globe and Mail* (25 February): pA16.

McKenna, B. (2015). Investing in Ontario's Economy – or Not. *The Globe and Mail* (5 December): pB1.

McLuhan, M. (1964).*Understanding Media: The Extensions of Man*. McGraw-Hill, New York.

Morgan, M. (2002) *Against the Mainstream; The Selected Works of George Gerbner*. Peter Lang Publishing, New York.

Picard, A. (2017) Children Don't Need a Daily Diet of Junk-Food Ads. *The Globe and Mail* (7 March).

Pope Francis. (2015) *Encyclical on Climate Change and Inequality: On Care for Our Common Home*. Melville House Publishing, London.

Postman, N. (1986) *Amusing Ourselves to Death*. Penguin, New York.

Rubin, J. (2015) *The Carbon Bubble*. Random House, New York.

Solyom, C. (2015) Extremism Runs Wild On The Web. *The National Post* (27 June): A14.

4 Funding policy choices

Tax and global financial secrecy

Michelle Gallant

Discrete constituencies differ on the particular attributes of a better tomorrow yet the realization of any collective vision depends upon resources. Secrecy hampers that realization by helping to hide billions from the dictates of public policy. It creates a dark universe that caches wealth from taxation, depriving states of moneys that might be spent on reducing child poverty, on tackling global warming, on easing the plight of refugees or on mitigating social and economic inequality. The public policy engine sputters, lacking the revenues needed to fuel the realization of collective aspirations.

In a modern world in which multiple compelling exigencies compete for attention, concern with secrecy and global tax-flight might seem quaint, the rarefied world of those preoccupied with the affairs of the wealthy at the expense of poverty, war and disease. Yet secrecy's abetting of the depletion of domestic treasuries relates directly to many, if not most, international problems. Wars are funded, in whole or in part, by state resources. The redistribution of tax moneys alleviates poverty and social and economic inequality. Health care is a primary destination of public funds. Catering as it does to tax-flight, secrecy undermines public choice. It relates directly to exigent concerns. It reduces public resources and thereby constrains public choices. Rather than quaint or the preoccupation of a rarefied world, secrecy goes right to the heart of the public's capacity to define and create its preferred vision of a just world. This paper explores secrecy's contribution to domestic revenue drought.

Secrecy and the global tax problem

Secrecy is a persistent theme in modern tax discourses. Untold wealth allegedly escapes taxation by choosing to reside in regions known for their strong protection of secrecy and for their favourable tax rates (Palan, Murphy and Chavagneux 2010). Vast amounts of untaxed earnings circulate in a cavernous 'subterranean' world of financial darkness (Christensen 2012). Secrecy dominates narratives on the exploitation of developing nations, associated with the stripping of tax moneys from domestic budgets (Oxfam 2016). Its influence in generating tax deficits culminated in the creation of a financial index, a device that ranks states on their levels of secrecy protection. No global tax discussion is bereft of an allusion to the problem of secrecy.

Secrecy reverberates in the global, rather than local, tax discourse because that is the forum within which it thrives. Were parochial local orders to constitute the entire matrix with which taxation operates, any frustration could be remedied by flexing of the powers of domestic governance. Tax shadows could be alleviated by their simple interdiction, subject, of course, to the constitutional constraints. Increasingly fluid global economic ordering has largely consigned insular domestic taxation environments to a utopian past. Although taxation remains housed in domestic law, foreign elements disrupt its operations. The discernment of corporate and individual income tax liability depends upon transactional activity that occurs in foreign jurisdictions. Business operates in a global setting, buying, manufacturing and selling goods around the globe. Individuals derive income from domestic and foreign sources. The determination of tax liability habitually involves consideration of the implications of transactions that happen beyond domestic boundaries. Shadows, whether the product of specific secrecy protections or of complex global financial arrangements, emerge from an interconnected global tax environment over which domestic legal authority has little influence. And, in truth, it is debatable whether taxation has ever been simply a local matter, or, at least, when taxation ceased to be a matter of simply local import. A vast exodus of wealth, often pressed by exigent domestic tax demands, has long plagued the realization of domestic revenue ambitions (Brion 1912).

It would be incomplete to explain the thriving of tax-related secrecy as simply an organic consequence of globalization. While it does arise from the decidedly more permeable nature of global affairs, it is equally something that is consciously cultivated. States have long intentionally legislated the enhanced protection of secrecy in order to attract global investment. Outright refusal to share of financial information, whether in a tax context or others, has long been the prevailing orthodoxy.

Nor is tax-related secrecy the exclusive preserve of nation states. It is constructed by the private sector, orchestrated by financial professionals who seize opportunities to minimize the tax burden. The modern tax universe is cluttered with sophisticated financial edifices. Modern wealth managers craft convoluted global financial mosaics which impede the detection of tax liability or impede the extraction of knowledge relevant to the attribution of domestic taxation liability.

In this, the world of secrecy is often confused with the world of tax crimes. Domestic law ordinarily recognizes a distinction, although often an indiscernible one, between the intentional deceitful evasion of tax norms and the intentional avoidance of tax norms. The former is a crime: the latter is not. Elaborate global financial structuring animates both contexts. The seizure of opportunities to minimize the tax-bite, short of the commission of a crime, has long been acknowledged as legitimate. Tax law admits the privilege of arranging one's affairs to attract the lowest tax burden (*Duke of Westminster v. Inland Revenue Commissioners*, [1936] AC 1.). The courting of secrecy by tax professionals, when it engages with mere tax-planning, is not engagement in crime.

Eloquent global financial structuring is a significant contributor to domestic revenue drought. It is in the manufacture of that eloquence that secrecy assists. While it may not be criminal to orchestrate massive tax depletion, the complexity inherent in global tax-planning makes it difficult to discern whether unlawful tax evasion has occurred. It also makes it incredibly difficult for revenue authorities to trace and discern the tax implications of global edifices, particularly when edifices straddle two or more states with firm commitments to financial opacity.

A terse history: the Second World War to tax, crime and resistance

The alliance of bank secrecy with the Second World War is infamous but had little to do with the dodging of domestic tax norms. Somewhat less infamous is the mid- to late-twentieth-century propagation of secrecy and its specific fusion with taxation or its latter-day captivation by the money-laundering discourse. The Second World War alliance sparked revulsion. The less infamous developments sparked concerted efforts to dismantle secrecy.

Amid the overarching atrocities that animated the Second World War, the War underscored at least two financial realities: it cast a deadly pallor onto any notion of secrecy; and it demonstrated the value of discretion in financial affairs.

The Second World War sowed distain for the financial secrecy. Swiss bank secrecy fuelled the German war effort and concealed the spoils of war – including the vile theft of the assets of the victims of Nazi horrors (Ziegler 1997). Post-war, persistent anonymity impeded efforts to recover the assets of Holocaust victims and others (Orland 2010).

Cataclysmic in its implications, this episode shoved secrecy out of the nether regions of banking and finance to which it had largely been confined and into the public domain. Proverbial references to Swiss financial secrecy have infused popular culture ever since. The classic Swiss defence of its financial role in the war – that a retraction of secrecy might have attracted an invasion and that Switzerland was surrounded by a powerful potential adversary – has some merit. However, the solid walls remained after the war when the threat was no longer present. Equally, Switzerland long avowed that its domestic secrecy laws were reinforced to help the victims of confiscatory German laws hide their resources. This defence has largely been consigned to cultural myth.

The shadows that shaped the Second World War travesties were explicit creatures of Swiss law. The disclosure of financial information was, under domestic law, a criminal offence. To open the vaults to scrutiny was to invite domestic prosecution. Together with this broad injunction, national law enabled the infamous anonymity vehicle, the numbered Swiss bank account. Numbered accounts sheltered the identity of account holders by identifying account holders simply by number rather than by name. When the war ended, this tool proved damagingly instrumental in denying the claims of Holocaust victims (Orland 2010). Inheritances could not be traced by family names, only by a number.

In any secrecy narrative, this epoch stands as a glaring indictment: it revealed the vile capacity of financial shadows to agitate evil. Axiomatically, it underscored anonymity's attractiveness. Bank secrecy brought bullion to Switzerland. The Helvetic state emerged from the war a powerful international financial centre, a position it has never surrendered. On the one, the reputation of secrecy was indelibly stained. On the other, discretion was shown to be a valuable commodity.

The evolution of secrecy proceeded more tacitly from the Second World War onwards. From the mid-twentieth century onwards, a less disturbing alliance propagated and attracted virtually no wide public interest. States perceived the economic benefits of financial discretion and incorporated this into developmental strategies. Anonymity flourished not amid any explicit couplings with war but under the explicit coupling of financial secrecy with offerings of low taxation rates.

The Caribbean islands were among some of the first to pioneer this coupling (Brittain-Catlin 2006). To solicit foreign investment, overt guarantees of financial discretion were fused with tax advantages. Other jurisdictions were quick to endorse this strategy (Hampton 1996). The pattern prospered in areas that latter came to be somewhat pejoratively defined as 'Off-shore'. The ensuring decades witnessed the vast expansion of the financial services industry, the building of domestic empires that serviced finance on a global scale. These empires were not forged exclusively from the twinning of tax with tight privacy rules – stability and proximity to existing financial centres were influential (ibid.). But the twinning proved a ubiquitous centrepiece.

As a developmental initiative, this strategy appealed because it did not depend upon any natural endowments – such as access to oil – and was relatively cheap to implement. It required the political will to enact norms that facilitated exchange and an educated elite, domestic or imported, to realize those advantages.

Under burgeoning globalization, this commodification of secrecy influenced financial trajectories. Attracted by privacy and meagre tax demands, international corporate empires migrated their taxable earnings. Wealthy individuals reduced their domestic tax burdens by shifting their residency, a classic basis of liability to taxation, to financial havens. Resources aggregated in these havens rather than accumulate in places with substantial connections to the production of that wealth. This century long propagation, the solicitation of foreign moneys by tying secrecy with low tax rates, underpins contemporary global tax tensions.

In tandem with discrete state-based solicitations of secrecy, the sophistication of global finance contributed its part to financial opacity. International financial structuring, including reliance on complex corporate arrangements, began to be regularly deployed to relocate taxable earnings to favourable tax climates. Discrete financial vehicles that serviced both global mobility and anonymity appeared – including trusts, which had long helped secure certain local tax advantages, and 'shell corporations'. The particular vehicle of the cross-border trust, which severs the beneficiaries of wealth from the legal owners, swiftly became a normal feature of international finance (Harrington 2016). Shell corporations, omnipresent creatures of international finance, are entities incorporated in jurisdictions in which they have no significant underlying economic attachment. Tax opacity matured through the layering of discrete corporate vehicles one onto the other, through the use of trusts, shell corporations and other vehicles to route financial activity through multiple jurisdictions. Sophistication and international complexity bread tax obscurity. Over the course of time, the tax world morphed into a dark complex universe into which untold wealth escapes.

The most recent twist on the narrative is less the maturing of secrecy than its captivation by the money laundering discourse. Over the last few decades, havens of privacy, vehicles of anonymity and complex global financial machinations merged with money laundering, the idea that the dark world was awash with money linked to, or derived from, criminal activity (Gilmore 2011). That this money took refuge in the world of shadows was blamed for fomenting the drugs trade, corruption, terrorism and other serious crimes. This prompted the demonization of money laundering, the act commonly defined as concealing, or 'cleansing' resources linked to crime. This enveloped the secrecy discourse in more than simply tax concerns. It is the anonymity of criminal money, together with the lack of transparency of the tax cosmos, that sparked attempts to dismantle secrecy.

Dismantling secrecy: transparency initiatives

This grim story of the Second World War prompted attempts to pry open Swiss secrecy but admitted little success. The gradual awakening to the implications of globalization on taxation stirred a broader discourse on international tax, of which secrecy dismantlement forms a part. The problem of money and crime catalysed a global assault on money laundering, the central objective of which is to increase the transparency of the global financial machinery.

Tax and secrecy dismantlement

While the discourse on international tax is not a discourse on secrecy *per se*, dismantling financial shadows proves a central focus. Initially, the international community was concerned with double-taxation, globalization's birth of the conundrum of the attraction of tax liability in more than one jurisdiction. To date, the only instrument on taxation that occurs under the auspices of the United Nations is a model treaty on double-taxation (United Nations 2011). As the devices and practices of secrecy stretched, tax transparency assumed greater paramountcy.

In 1985, the Organisation of Economic Co-operation and Development (OECD), the organ that has assumed principal responsibility for managing the global tax discourse, released a report that identified the perilous link between bank secrecy and tax-flight. In the 1990s, the OECD evoked controversy with its identification of 'tax havens' as problematic phenomenon (OECD 1998) Controversy centred less on the fact that 'tax havens' were havens of secrecy and centres of tax depletion but on the membership in this demonized club (Sharman 2006). Scrutiny fell more fully on the practices of more marginalized, less powerful, states than on the potential 'tax haven' status of more influential ones.

Demonization of tax havens gradually ceded to broad consensus on for greater tax openness and the global sharing of information related to taxation matters. Global tax-transparency standards, the automatic exchange of certain tax-related information and the standardization of global accounting standards gained credence and are beginning to achieve some degree of harmonization (OECD 2015). A global forum emerged to press these transparency standards, rather than leave compliance to the whims of individual states (OECD 2017).

Of course, transparency norms form only a piece of efforts to stem the problem of the tensions between a global economic order and domestic tax orders. It is a crucial piece, however, because transparency renders apparent and amenable to public scrutiny aspects of revenue drought that would otherwise lies embedded in shadows.

Crime and secrecy dismantlement

Slightly older, and more formalized, is the broader assault on secrecy that derives from the money laundering discourse. The links between shadows and crime congealed into four international treaties, the latest some ten years old. Cognizant that compliance with international undertakings often matures at a glacial pace, a specific organ, the Financial Action Task Force (FATF), was convened to push transparency efforts. Under its influence, secrecy dismantlement in this context has moved an impressive pace.

Drawn principally from international treaties, the standards released by the FAFT are known as the international anti-money laundering standards. This formalized edifice speaks directly to secrecy and, more generally, to international financial transparency.

The FATF standards hold that no state refuse to implement anti-money laundering law on the basis that that action would violate domestic secrecy protections (FATF 2012). As noted earlier, Swiss domestic law criminalized the disclosure of financial information. The extant standard denies any privileging of secrecy, or any failure to comply with the international framework on the basis of any prioritizing of domestic law.

Less directly, the global apparatus seeks to build a legal culture of transparency. Financial intermediaries, banks, professionals and others whose work touches on finance – are required to report any suspicious financial activities – including transactional activity potentially linked to the crime of tax evasion. Intermediaries must completely identify the clients of financial services and maintain detailed records of financial transactions – including the proper and full identification of the ultimate beneficiaries of trust agreements. The standards specifically deny the legitimacy of shell corporations and require that financial service providers establish internal anti-money laundering programs.

Encompassing in its remit, global anti-money laundering regulation attends most directly to crime. Tax evasion fits directly into this narrative. While the specific inclusion of tax-crimes into the global apparatus was extremely late to arrive, tax offences are formally listed as offences coming within the money laundering fold.

Confronting a global culture of resistance

Transparency developments hold considerable promise. Had express denunciations of secrecy been formalized decades earlier, assistance to the German war effort might have been forestalled or the commodification of anonymity restrained. To engage with secrecy, however, is to engage with power. Secrecy protects wealth: it keeps resources in private hands. The depth of commitments to financial anonymity and the circumvention of tax norms can be stunning. Empirical glimpses of this world can be elusive. And the interest in protecting wealth from avaricious public tax laws may adapt to transparency initiatives by finding new ways to keep tithings in private hands.

United States Senate inquiries in the early days of the twenty-first century proffered a rare glimpse of the actual servicing of international tax-dodging (United States Senate 2014). Whistle-blower Bradley Birkenfeld's disclosures of information about the internal workings of a financial institution revealed stunning institutional commitments to the dodging of domestic tax norms. The revelations revealed a veritable corporate culture of tax evasion. Internal corporate communications, otherwise privileged, offered specific guidance on how to solicit US wealth while simultaneously avoiding the reach of United States tax norms. The attendance of foreign bankers at events on US terrain likely to attract high-net worth US citizens were coupled with express warnings not to conclude any business on US territorial terrain. Such conduct would attract US tax jurisdiction. Moreover, the attendees at these high-net worth events regularly noted on their customs declarations – the documents signed to secure entry into the United States – that they were in the country for the purposes of pleasure, not business.

Such formidable glimpses of the world of financial darkness are rare. This is because privacy constraints shield information. Birkenfeld was a whistle-blower, a banker who broke ranks and shared his secrets. Any investigation of the world of secret finance is denied by its own rules. Financial secrecy is the act of not sharing information, of refusing to disclose.

A more recent glimpse of the universe was provided by the International Consortium of Journalists. Massive investigations by a committed group of journalists resulted in the release of information that substantiated the existence of vast hidden economy (see www.panamapapers.icij.org). The release, or theft, of the information arguably breached the norms of financial privilege. Consolidated in the form of a comprehensive database, the materials disclose the trajectories of finance, the reliance on trusts, on corporate entities and on multiple inter-jurisdictional transactions to reduce the tax bite (Cockfield 2016).

Unorthodox and irregular, the work of the journalists and the US Senate investigations suggest the continued vitality of the secrecy industry. Both speak of the problem of penetrating any shadow universe. The very information needed to analyse any global tax-flight is buried beneath the cover of secrecy. And without firm data, without knowledge, it is difficult to prescribe remedies.

Transparency challenges the secrecy culture. Cultures are known to adapt to exigent circumstances. France's recent embrace of greater tax transparency arguably prompted some of France's wealthiest citizens to shift their earnings from European financial institutions to less stringent regulatory climes (Zucman and Johannesen 2014). Rather than surrender any booty to the taxman, taxable earning may merely continue to migrate. Some early study of nascent rules indicates that the path of wealth shifts under the force of transparency (ibid.). In this, in confronting the wealth's disinclination to contribute to public coffers, the culture of secrecy adapts.

Conclusion

A better tomorrow requires its visualization and its realization. Realization depends upon resources and secrecy denies access to billions in tax dollars. Crafting innovation solutions to complex ecological problems may very well be costly but the choice to allocate tax dollars to such imperatives lies with the public. Secrecy does not provide adequate and safe food. Secrecy does not clean the environment. Secrecy does not promote sustainable growth. Secrecy disempowers the collective. To enable public choice, to realize the collective vision of a better tomorrow, secrecy must recede.

References

Brion, Maurice, 1912, L'Exode des Capitaux Francais a L'Etranger, doctoral thesis, University de Nancy.

Brittain-Catlin, William, 2006, *Offshore: The Dark Side of the Global Economy*, Picador, New York.

Christensen, John, 2012, 'The Hidden Trillions: Secrecy, Corruption and the Offshore Interface', *Crime, Law and Social Change*, 57(3), 325–343.

Cockfield, Arthur, 2016, 'Big Data and Tax Haven Secrecy', *Florida Tax Review*, 18(8), 483–539.

FATF, 2012, *International Standards on Combatting Money Laundering and the Financing of Terrorism and Proliferation*, Financial Action Task Force, Paris.

Gilmore, William, 2011, *Dirty Money: The Evolution of International Measures to Counter Money Laundering and Terrorist Financing*, Council of Europe, Strasbourg.

Hampton, Mark, 1996, *The Offshore Interface: Tax Havens in the Global Economy*, St Martins Press, New York.

Harrington, Brooke, 2016, 'Trusts and Financialization', *Socio-Economic Review*, http://works.bepress.com/brooke_harrington/44.

OECD, 1998, *Harmful Tax Competition: An Emerging Global Issue*, Organisation for Economic Co-operation and Development, Geneva.

OECD, 2015, *Countering Harmful Tax Practices More Effectively, Taking into Account Transparency and Substance, Action 5-2015 Final Report*, Organisation for Economic Co-operation and Development, Geneva.

OECD, 2017, 'Global Forum on Transparency and Exchange of Information for Tax Purposes', www.oecd.org/tax/transparency.

Orland, Leonard, 2010, *A Final Accounting: Holocaust Survivors and Swiss Banks*, Carolina Academic Press, Durham, North Carolina.

Oxfam, 2016, 'Tax Battles: The Dangerous Global Race to the Bottom on Corporate Tax', December, www.oxfam.org/sites/www.oxfam.org/files/bp-race-to-bottom-corporate-tax-121216-en.pdf.

Palan, Ronen, Richard Murphy and Christian Chavagneux, 2010, *Tax Havens: How Globalization Really Works*, Cornell University Press, Ithaca, New York.

Sharman, Jason, 2006, *Havens in a Storm: The Struggle for Global Tax Regulation*, Cornell University Press, Ithaca, New York.

United Nations, 2011, 'Model Double Taxation Convention Between Developing and Developed Countries', www.un.org/esa/ffd/documents/UN_Model_2011_Update.pdf

United States Senate, 2014, *Offshore Tax Evasion: The Effort to Collect Unpaid Taxes on Billions Hidden in Offshore Accounts*, Permanent Subcommittee on Investigations, Committee on Homeland Security and Governmental Affairs, 26 February, United States Senate, Washington, DC.

Ziegler, Jean, 1997, *The Swiss, the Gold and the Dead: How Swiss Bankers Helped Finance the Nazi War Machine*, Harcourt Brace, New York.

Zucman, Gabriel and Niels Johannesen, 2014, 'The End of Bank Secrecy? An Evaluation of the G20 Tax Haven Crackdown', *American Economic Journal: Economic Policy*, 6(1), 65–91.

5 Bruno Latour on ecology and Christian religion

Philippe Crabbé

> What risk do we run? ... The world is young, the sciences are recent, history has barely begun, and as for ecology, it is barely in its infancy. Why should we have finished exploring the institutions of public life?
>
> (Bruno Latour, *Politics of Nature*, quoted by De Vries 2016: 157)

For a Christian environmentalist, Bruno Latour's views on ecology and Christian religion are mutually reinforcing. Latour's most recent book, *Face à Gaïa* (2015; English translation published as *Facing Gaia* in 2017), deals with the current ecological crisis, mainly climate change, and its remedy. The remedy proposed by Latour is political and requires a complete representation of agencies (i.e. abilities of any object to change a state) to which territories and, therefore, sovereignties are assigned.

Latour is a catholic who loves his religion ('A person of faith has only one religion, as a child has only one mother'; Latour 2005: 28) as witnessed by his older book *Rejoicing: Or the Torments of Religious Speech* (2013; French original 2002). However, Latour feels that Christian religion is in need of renewal: 'It is from this very impossibility of speaking to my friends and to my own kin about a religion that matters to me, that I want to start... Religion, in my tradition, in my corner of the world, has become impossible to enunciate' (ibid.). 'Christians take as proof of the tediousness and decadence of this age what is in fact the result of their own laziness in pursuing the translation[1] task of their fathers' (Latour 2001: 228).

The connection between *Facing Gaia* and *Rejoicing* is the following:

> religion, in its Christian instantiation at least, presents itself as a rather plausible alternative to an ecological consciousness whose ethical and emotional drives do not seem to have enough petrol (or soybeans) to carry us through the tasks it has burdened upon us ... we should not forget that the appeal to renewing everything, here and now, and in this world, is first of all a religious passion ...
>
> (Latour 2009: 463)

Besides being 'among the most important figures in contemporary philosophy and social science, according to philosopher of science George De Vries (2016),

Latour has studied the production of natural sciences extensively (one of his fields of research was science studies; he calls himself an anthropologist of science and technology) and is at ease almost as much within the Anglo-Saxon culture as he is in his own French one. He is a superb communicator as well. His ethnographic studies have revolutionized our understanding of areas as diverse as science, law, politics and religion. Latour has formulated not only the philosophical groundwork for an alternative worldview, but has offered also the means to further empirically explore the world' (De Vries 2016: 156). To facilitate a more realistic understanding of the world, Latour has introduced a radically fresh philosophical terminology and a new approach to social science (actor–network theory). This approach emphasizes relations between things and humans and is, therefore, a close kin to ecology.[2] Moreover, Latour, lives in a post-Darwinian world which animates nature (Miller 2013). Each object is on its own to alter its environment. However, Latour distrusts wholes. There is no macro-force in his worldview. Latour has put all the factors involved in a social situation on the same level and has done away with the concept of social forces. Thus, objects, ideas, processes, and any other relevant factor are seen as just as important in creating social situations as humans. The actor–network theory can more technically be described as a 'material-semiotic' method. This means that the latter maps relations that are simultaneously material (between things and people) and semiotic (between concepts) (De Vries 2016).

While Latour' s ecological thinking is pretty much self-contained in *Facing Gaia*, this is not the case for his religious thinking; *Rejoicing*'s French original dates from 2002 and must be completed by his articles on religion (see bibliographical references). While both George De Vries (2016) and Adam S. Miller (2013) have ably synthesized Latour's thinking, Miller devotes more space to Latour's views on religion, God and metaphysics than does De Vries.

Facing Gaia

Latour finds in James Lovelock a justification for his actor–network theory and in Michel Serres for elements for his political solution to the mutation provoked by the advent of the Anthropocene. The latter allegedly challenges modernity. While Hobbes started from a state of nature to elaborate his social contract, we have to start from an acknowledged state of war among all agencies to establish new sovereignties for the latter. The solution proposed is a political application of what economists call Coase theorem (Coase 1961).

The two-fold point of departure of the book

The book *Facing Gaia* has a twofold point of departure. The first is the 'mental image' of Gaia, the earth and its atmosphere (Lovelock 2000). Latour insists that Gaia is not a cybernetic system with a 'governor' but rather a set of actions and reactions without hierarchy and harmony. Gaia is neither a 'mother', nor a system but a history within which there are neither subjects nor objects. In this

Latour's sociological mentor is more Gabriel Tarde (1843–1904) than Émile Durkheim (1858–1917).[3]

The second point of departure of *Facing Gaia* is the natural contract of Michel Serres (1995; French original 1990), which transposes Hobbes's social contract into a contract between human interest groups and agencies from the balance of nature. For Latour, being Darwinian, agency is no longer humans' exclusive prerogative. Since the balance of nature is unable to negotiate, it is through humans that it must be represented. Only those endowed with agency are entitled to speak within this representation of the balance of nature. Thus nature has a plural voice, is disaggregated into several agencies; nature, as a unit, never existed according to Latour. This was not the case for Serres and this plurality of agencies is one of Latour's main innovations.

The mutation of the Anthropocene

We are not experiencing a passing ecological crisis but a mutation forever, the Anthropocene, i.e. an age in which humans are able to leave geological traces. What used to be considered as external decor, the 'environment' has become an 'actant' internal to history.[4] If the crisis is really a mutation, it should have been perceived as such by humans but it has not been. We have become those who could have prevented the mutation but did not. We went through a total and colonial war on the 'environment' without noticing it. The precautionary principle has not been respected except in relatively trivial situations close to us. The problem is behind us. Modernity has normalized danger. We have become quietists (God, technological progress will take care of it all!). At the other extreme, geo-engineering wants to push modernity to its limits. Others want simply to apply existing solutions and institutions to the mutation but this won't work.

Latour rejects the dualism nature/culture because, in Gaia, nature is no longer perceived as inert. Streams and chemicals are endowed with agency and science multiplies agencies. Nature is not separated from humans any longer. Since nature is not inert, since there is no distinction between nature and culture, modernity has ended. The Anthropocene has also ended anthropocentrism and any attempt at unifying mankind.

The climate science

Climate science is not founded on theoretical physics or mathematics but on a consensus among various empirical sciences. Negationists are no longer marginals challenging elites but elites challenging other elites. The dualism description/prescription is non-existent because the latter two are actually strongly coupled; description is never neutral. When you describe, you also lead to actions. Therefore, there is no distinction between science and politics any longer. Ideology – mixing facts and values – is harder to detect. It is not the science but the scientific institutions one needs to protect. One needs political power for the scientific institutions because climate science does not speak with any authority.[5]

Thinking spherically

Following Peter Sloterdijk (2010; German original 1999), we build a protection membrane, a sphere, a globe around ourselves. We try to jump from the element level to the level of wholes as if elements were parts. And we claim that the whole is superior to the sum of the parts. We confuse whole with what is more connected. Scale is not holarchic (i.e. the larger containing the smaller). Scale simply means that a larger scale is more connected than a smaller one. The globe is a consequence of Platonic philosophy, which has been used in theology. This is why there is no history, no time, no space in medieval Christian theology. The same can be said for scientific reason. Epistemology is of the nature of a globe as well. Thinking globally is impossible unless one talks of a reduced model of reality. Scientific reason needs to be situated. Gaia has been the subject of globe construction as well by Lovelock's critics (i.e. to a process which confuses the whole with the connexions among the elements). Lovelock did not globalize, did not recognize two levels. Humanity, nature, God are other examples of these global concepts. For example, humans are not unified into humanity, since individuals are subject to different feedbacks and, therefore, to different responsibilities (e.g. nations do not speak for ecological functions).

Modernity and the challenge of the Anthropocene

Modernity is characterized by an inanimate nature, a world governed mechanically from outside by the laws of nature, a scientific truth as solid as geometric propositions, people governed by self-interest, a literal interpretation of the Bible (Counter-Reformation).

Contrary to modernity, the Anthropocene animates nature with agencies. It is subject to neither exterior nor interior authority. Religion has not disappeared as predicted by modernity and science is not separate from religion as modernity once intimated.

Cosmology and the translation tables methodology

A substance is defined by its attributes. If the attributes are the same, the substance must be the same even if the names of the attributes differ (e.g. the national gods in pantheism). A cosmology is defined by five parameters: a supreme authority, limits to its people (who is included?), a territory, a time period to which the cosmology applies and an organization principle ('cosmogram').

A natural science or a religion may be defined by its epistemology (what it claims to be doing) or by its anthropology (what it actually does). The four combinations may be compared. Using the translation tables ethnological methodology (due to Jan Assman 2003; see Table 5.1 below), Latour argues modernity has simply substituted another deity or supreme authority (Laws of Nature instead of God) and another territory (or rather an absence of territory in the case of science to another world in the case of religion) to the medieval

Table 5.1 Translation tables ethnological methodology.

	Natural religions (epistemological)		*Terrestrialization (anthropological)*	
	Nature 1	*Counter-religion 1*	*Nature 2 (critical)*	*Counter-religion 2*
Deity	Laws of nature	Ordering God	Multiverse (Gaia)	God of ends/end of ordering God
Cosmogram	exterior	exterior	interior	Local (no universal)
	unified	unified	multiple	multiple
	Dis-animated	Over-animated	Animated (agencies)	animated
	unquestionable	unquestionable	controversial	controversial
People	everybody	everybody	scientists	Church
Territory	Extra-territorial	From another world	Connected to networks	Incarnated (no escape from this world)
Epoch	Radical break-up	Radical break-up	Multiple temporality	Constantly new beginnings

Latour (2015: 232), © Editions La Découverte, Paris

cosmology. The replacement of God by evolution, nature or Gaia does not change anything as long as the attributes remain the same: deity and nature remain connected. This guarantees civil peace. Pantheism used to connect people while monotheism has connected religion and truth. Since the advent of monotheism, religion no longer reconnects. Monotheism has thus become a counter-religion (Assman 2003).

The assassination of Henri IV of France ended the period of tolerance, both political and in the realm of ideas, which existed during the Renaissance. The scientific counter-revolution or modernity, which, like monotheism, associated knowledge and truth, is, therefore, deeply religious and gnostic. This is why the warnings of climate science have not been heeded. If you don't like the prescriptions resulting from science, you question the science. The moderns believe they can realize the end of time in history, i.e. the virtual realization of paradise on earth.

The new 'nomos'

The question we have to face is how we can turn this collective (a pluriverse of actions and reactions) into a common one. Plurality generates conflict that we try to solve by a commonly acknowledged authority but there is no commonly acknowledged authority any longer. There is no unified nature (such as the reference to the latter in natural law) to which we can refer in order to resolve conflicts; there is only a history which is the one of the relationships within Gaia.

In this sense, following Carl Schmitt (2001; German original 1950), all agencies are at war with each other and the war deals with territories bound by borders, with drawn limits to overlapping interests but at variance with national ones. The territory is endogenous to history. The nomos is an immediate configuration under which the social and political order of a people can be spatially perceptible. It is equivalent to the distribution of agencies, the 'cosmogram'. The nomos or cosmogram requires the acknowledgement that we are in a war situation. The enemy is the alien, including the non-humans (i.e. the one with whom conflicts can't be resolved by appealing to general norms established in advance or by an umpire). Territories can no longer be extended by the addition of new territories or by new scientific discoveries. To use economic terminology, we have reached the limit of the extensive margin. Therefore, we need to work at the intensive margin only (i.e. redistribute territories through war). We have reached the great enclosure. Peace can now be achieved through politics only.

Science is incapable of playing the role of umpire any longer. Scientists are endowed with a territory as well, the one defined by their instruments. However, the former are unable to influence the climate change political debate because all the interests or agencies endowed with territories, Carl Schmitt's nomos, are not represented in the debate. Economists call this phenomenon, market failure. On the other hand, other agencies do not believe they are endowed with a territory; this ignorance renders them indifferent to the consequences of their actions. This is the case for modernity. Scientific institutions therefore need political power. Following Carl Schmitt, it is the non-recognition of the enemy which turns limited wars into extermination wars. This is the case with ecology. It is by not acknowledging non-humans as enemies capable of wiping us out that the current situation has turned into an annihilation war.

Stewardship, sustainable development, reengineering are all pacifying operations which by-pass politics. All of them rely all on a unifying nature capable of conflict resolution. One has to assign territories to non-humans, the new 'nomos': one has to re-politicize ecology. NGOs are questioning territories that countries believe they alone possess. The list of one's interests is extended by negotiations; it is difficult to assign limits to one's interests under Gaia. It's by assigning territorial limits that one decides the distribution of agencies, the cosmogram. Moderns are in a hurry to escape the territories of the past but indifferent to the territories they are going into except through utopias; they are focusing not on the future but on the past.

Economic critique

'If the whole of Lovelock's cunning, as I have sufficiently showed, consists in disaggregating the system in as many actors capable of overlapping the actions of others, it is this dis-aggregation of agencies that one needs to translate politically in order for the territories' overlaps to become clearly visible' (Latour 2017: 342).[6] This process constitutes effectively a spatial translation of the economic concept of externality or of the commons. The political solution offered by

Latour is actually a special case of the one offered by Ronald Coase in the 1960s (i.e. bargaining among all the parties to an externality). This is what Latour calls the distribution of agencies, i.e. the sharing of sovereignty. He suggests an historical analogy, Pope Gelasius's theory (ad492–496) of the two powers, the spiritual one and the temporal one which overlap. Latour illustrates his solution through national institutions which are endowed with quasi-exclusive jurisdiction over certain resources such as water.

Ecological critique

The assertion that Gaia is not a system is hardly acceptable to ecologists who have studied hierarchical systems in depth. The ecological concept of hierarchy is constructivist (Ahl and Allen 1996) as much as Latour's Actor-Network Theory. However, for Latour, science is always mediated, situated and historical (Latour 1998).

Hierarchy theory focuses in finding a powerful point of view rather than on what is happening in the real world (Ahl and Allen 1996). Hierarchy does not correspond to an objective scale but rather to a constructed spatio-temporal scale where phenomena occur at several scales. If agencies are empirical entities, the former must be ordered according to their scales of activity. Implicitly while accepting that the earth went from the Holocene to the Anthropocene, Latour admits that human influence affects the top of the spatio-temporal hierarchy (i.e. the geological ages at the scale of the planet). When Latour talks of Gaia, he doesn't talk about what Ahl and Allen call a simple large numbers system in which the behavior of lower level agencies dissipates in the average condition of upper level emergent properties. When Latour talks about 'forest' or 'ocean' or 'water', he talks of aggregates of connected elements, and thus of subsets of the set of agencies, of contexts which are endowed with territories (i.e. spatial scales), and which are endowed with temporal scales as well since agencies function with different frequencies. Agencies must have characteristic scales (i.e. a typical extent over which a process is expressed, e.g. seasonal). Therefore, agencies form hierarchical systems even if their levels are observer determined.

Structure is a call of judgement from the observer in hierarchy theory. Relation between structure and scale is at the heart of it. While our experiences of the world are linked to change over time (dynamics), our definitions which correlate with patterns of repeated experience (structure), are static and available to serve as elements in cognitive models. When a label is attached to a pattern, the effect is to make the dynamic become static (Ahl and Allen 1996). Latour calls this pattern 'freeze-frame' and distrusts stable hierarchies while acknowledging that something has to remain constant when we transform information through different media, different representational states. And this something he calls 'immutable mobiles' because 'information is never transported without being deeply transformed' (Latour 1998: 429).

Large scale is the context, such as the Anthropocene. Issues of scale enter an investigation as soon as a measurement regime is established. A measurement protocol defines the scope of observation by establishing temporal and spatial limits (Ahl and Allen 1996). For Latour, scale is defined by the number of connections. But isn't the number (and directions) of connections more a characteristic of interdependence (indecomposability) than it is of scale? If Gaia is history, agencies and connections among them must be observable and act at different spatio-temporal scales according to the measurement regimes selected. This is not elaborated by Latour at least in *Facing Gaia*. More importantly, for Latour, the problem is not to adopt 'a powerful point of view' but to look at the world itself.

While the book is self-contained and well-written, *Facing Gaia* constitutes difficult reading because it is a synthesis of and builds upon previous works – not always easy to read despite their overlaps – in which Latour developed his constructivist methodology.

'Rejoicing'

Though, Latour considers Christian religion as a truth generator that he contrasts with others such as science and art, allotted space prevents this author to do full justice to Latour's understanding of religion. In a sense, Latour's understanding is not that new but his way of going about it is. Latour has a deep respect for Christian religion as long as the latter does not try to open up an immediate access to God (who needs to be mediated) or to convey knowledge and, thereby, compete with the sciences (e.g. creationism). Religion should not point to a transcendent beyond but rather to an immanent present that it should renew. Religion re-presents persons and repeats past things with a new tone. 'Meaning is always lost if it is not renewed now' (Latour 1998: 435).[7] Religion, rightly understood, is thus more capable to motivate people in reaching for ecological peace through politics than science or art does.[8]

God must be mediated

Religion is not superior to art or science. Neither is it superior to political talk in which the objective is to address a common identity. Religion must provide 'an obvious framework for ordinary everyday existence' in such a way that people understand religion in the same way they understand shopping. Like science, religion leads to an invisible reality. However, where science reveals transcendent objects by correcting for our myopia (we are unable to understand what's farther), religion reveals immanent objects by correcting for our hyperopia (we are unable to understand what's near us). The same competencies needed to be good at science are those needed to be good at religion. 'God' must be translated,[9] an act that Latour understands to be both transformation and renewal. The reality reached by religion is not distant, hard to understand but simply hard to renew.

Religion does not transport information but re-presents persons

It is not the transport of information but of persons, transformation which occurs in religion. This transport of persons is analogous to love talk. In a love relationship, the sentence 'I love you' is a gift and this gift generates personhood in the giver as well as in the receiver of the gift. Both are present to one another again and anew in 'the immediacy of assent' (A. N. Whitehead), the key to the power of religion (Latour 2001: 228). The conditions of this re-presentation are difficult to enunciate but easy to experience.

Conclusion

While Latour' s conditions of felicity – not detailed here – for Christian religion to succeed are not that new, they are presented anew as being fragile and deserving careful implementation. Latour' s metaphysics and understanding of God exclude the possibility of a traditional, omnipotent, wholly transcendent God who created the world out of nothing. God is just another mediation, another way of saying what is present and immanent (Latour 1998: 414). Latour's religion is strongly influenced by process philosophy (A. N. Whitehead), pragmatism (W. James) and Darwinism. However, through its passion, it may be able to renew contextual theology to which it is somewhat similar and, therefore, ecological consciousness. In my opinion, Latour's religious tone is truly capable of renewing Christian religion.

Notes

1 Translation in Latour's vocabulary means any change of a state into another one by any enunciation or performance which has an effect.
2 No wonder Latour is in awe of 'Laudato si', the current pope's encyclical on ecology, especially for its emphatic assertion that the entire creation will be saved.
3 'Here lies the real importance and radicalness of Latour's philosophy. Latour has moved pluralism from the level of cultures to ontology, to the plurality not of how we may see the world but to the world itself' (De Vries 2016: 154).
4 The word 'actor' is reserved to humans in Latour's terminology.
5 See Latour (2014) on this as well.
6 The page reference is to the French edition.
7 Though all this section is the gist of *Rejoicing*, I find that Latour (1998, 2001, 2005, 2009) are the most precise expressions of its content.
8 Latour, distrusting wholes, speaks of the Christian religion only that he knows.
9 See note 1 above for the meaning of Latour's translation.

References

Ahl, Valerie and Timothy F. H. Allen, 1996, *Hierarchy Theory: a Vision, Vocabulary, and Epistemology*, New York: Columbia University Press.

Assman, Jan, 2003, *Le Prix du monothéisme*, Paris: Aubier.

Coase, Ronald H., 1961, The Problem of Social Cost, *Journal of Law and Economics* 3: 1–44.

De Vries, George, 2016, *Bruno Latour*, London: Polity.

Latour, Bruno, 1998, How to Be Iconophilic in Art, Science and Religion? In *Picturing Science, Producing Art*, ed. Caroline A. Jones and Peter Galison, New York: Routledge.

Latour, Bruno, 2001, 'Thou Shalt Not Take the Lord's Name in Vain': Being a Sort of Sermon on the Hesitations in Religious Speech, *RES: Anthropology and Aesthetics* 39 (Spring): 215–234.

Latour, Bruno, 2005, 'Thou Shall Not Freeze-Frame', or How Not to Misunderstand the Science and Religion Debate, in *Science, Religion, and the Human Experience*, ed. James D. Proctor, New York: Oxford University Press.

Latour, Bruno, 2009, Will Non-Humans Be Saved? An Argument in Ecotheology. *Journal of the Royal Anthropological Institute* 15: 459–475.

Latour, Bruno, 2013, *Rejoicing: Or the Torments of Religious Speech* (tr. Julie Rose), London: Polity (French original 2002).

Latour, Bruno, 2013, *Facing Gaia: An Inquiry into Natural Religion*, Gifford Lectures, www.giffordlectures.org/lectures/facing-gaia-new-enquiry-natural-religion.

Latour, Bruno, 2014, War and Peace in an Age of Ecological Conflicts, *Revue juridique de l'environnement* 39(10): 51–63.

Latour, Bruno, 2015, *Face à Gaïa: huit conférences sur le nouveau régime climatique*, Paris: Editions La Découverte.

Latour, Bruno, 2017, *Facing Gaia: Eight Lectures on the New Climatic Regime*, London: Polity.

Lovelock, James, 2000, *Homage to Gaia: The Life of an Independent Scientist*, Oxford: Oxford University Press.

Miller, Adam S., 2013, *Speculative Grace: Bruno Latour and Object-Oriented Theology*, New York: Fordham University Press.

Schmitt, Carl, 2001, *Le Nomos de la terre dans le droit des gens du Ius Publicum Europaeum* (tr. L. Deroche-Gurcel), Paris: Presses Universitaires de France.

Serres, Michel, 1995, *The Natural Contract, Meditations on Environmental Change and the Necessity of a Pact between Earth and its Inhabitants* (tr. Elizabeth MacArthur and William Paulson) Ann Arbor, MI: University of Michigan Press (French original 1990).

Sloterdijk, Peter, 2010, *Sphères II*, (tr. Olivier Mannoni; German original 1999), Marensell.

Part II

Public health, environmental disasters and crimes against humanity

6 Reforming reparations for mass human rights abuses

A Canadian model

Kathleen Mahoney[*]

Introduction

This chapter will examine traditional domestic reparations for mass human rights violations and how they often fall short of what is required to make the victims whole. Traditional legal reparations for human rights violations are often unable to address the full range of harms caused by mass human rights violations and sometimes result in the re-victimization of those who were harmed the most. The chapter examines what is required to better serve victims of mass abuses, both procedurally and substantively. The role of the legal profession, law and the legal processes in the search for justice for victims is also examined. For the needed reform to take place, the legal, theoretical and policy frameworks of reparations for mass human rights violations need to be questioned and the Canadian model of the Indian Residential School Settlement Agreement (IRSSA) which created reparations for mass violations of human rights of indigenous Canadians, is a good example of the direction the law needs to take.

From the late 1800s to 1993 in Canada, mass human rights violations were committed against at least 150,000 indigenous children, their families and communities as a result of the imposition of the Indian Residential School policy.[1] Children were removed from their families and communities by the government in order to assimilate them into the body politic[2] with the view that the government 'could not kill the Indian but it could kill the Indian in the child.'[3] In 2005, the government of Canada settled claims of the 105,000 living survivors in the largest and most holistic class action settlement agreement in Canadian history and unique in the world. The settlement agreement was between Canada, the church entities hired to operate the schools, and the indigenous survivors who resided at the residential schools and their extended families.

The Canadian model is important for its approach, for its methodology as well as the substantive content of the agreement. It was a success[4] largely because it was sensitive to, and incorporated indigenous values and traditions in addition to going beyond the boundaries of domestic law to provide the kind of reparations that were meaningful to the survivors in ways the civil laws were not.

The legal right to reparations

In Canada, as in most countries of the world, victims of mass human rights abuses have a legal right to reparations. These rights are recognized in both international and domestic law.

International law guidelines state that victims of mass human rights violations have the right to prompt, adequate and effective reparations for gross violations of International Human Rights Law.[5] The UN Basic Principles and Guidelines provide a reparations framework described as follows:

a. Compensation to provide victims with monetary and nonmonetary damages to pay for the losses they have experienced;
b. Rehabilitation to repair the lasting damage of human rights violations through provision of medical, psychological, legal, and social services;
c. Restitution to restore the condition lost by the victim due to gross violations of human rights, such as the restoration of liberty, citizenship, employment, or property.
d. Satisfaction to cease continuing violations, disclose the truth, search for the disappeared or the remains of those killed, officially declare and apologize to restore the dignity, reputation, and rights of the victim, impose sanctions against perpetrators, and create commemorations and tributes to the victims.[6]
e. Guarantee non-repetition by initiating reforms to ensure independence of the judiciary, human rights education, mechanisms for preventing and monitoring conflicts, and reviewing and reforming laws and policies that contribute to gross violations of human rights.[7]

Survivors of Indian residential schools acquired rights to receive compensation because the injuries and harms inflicted on them at the hands of the perpetrators were human rights violations recognized in international and domestic human rights law and also recognized domestically as torts and crimes. In the Canadian case, indigenous children were arbitrarily removed from their homes and detained in the schools often for many years, if not their entire childhoods. They experienced injuries and poor health resulting from physical, sexual and psychological violence.[8] In their adult years they suffered lost opportunity, income and educational opportunities.[9] All of these harms are legally recognized harms in both international human rights law and domestic law and are compensable.

The right to rehabilitation, however, is not so protected. It arises when victims suffer harms that are ongoing and are in need of healing, counselling or treatment. Rehabilitation is partially recognized in domestic tort law when it is required by individuals, but when the harm extends beyond the individual to third parties, tort law does not recognize them.[10] In the residential school tragedy, the need for rehabilitation and healing reached beyond individual survivors to their families and future generations, who suffer from multiple layers of grief and pain from the loss of family life and loss of language and culture in the past and in the present.

Many survivors of residential schools explained that their trauma and dislocation negatively affected their parenting skills, and ability to form relationships and that their children and grandchildren were harmed as a result. These intergenerational harms are recognized in the international principles but remedies for them are non- existent domestically.

The right to satisfaction is similarly recognized in international law but not in domestic law. This right recognizes that in cases of mass human rights violations over a long period of time, the absence of judicial accountability should be compensated or repaired. In the residential schools, the lack of resolution about the fate of the missing children; the burial of the students who died in the schools and were buried in unmarked graves;[11] and their stigmatization as inferior, savages, stupid and unworthy human beings would entitle them to the remedy of satisfaction. Official apologies, truth commissions, provision for commemoration, healing funds, searches for the disappeared and memorialization are not remedies that can be sought in a domestic court of law.

The right to guarantee of non-repetition is another unrecognized right in domestic law but is recognized by the UN principles. Reforms to ensure independence of the judiciary, human rights education, and review and reform of laws and policies that contribute to gross violations of human rights cannot be accomplished by judicial remedies and require state action.

It is clear then, that the reparations outlined in international law are comprehensive and far more adapted to mass human rights violations than domestic laws are. The problem is accessing these remedies is a fundamental principle of international law that says in order to access international law, the injured parties must first exhaust domestic remedies. Even if domestic remedies are exhausted and a hearing is obtained before a relevant UN committee and succeeds, there are no mechanisms to enforce any remedies ordered other than persuasion, shame or diplomacy. Furthermore, the guidelines themselves are not legally binding. They are in the form of a resolution adopted by the General Assembly,[12] a non-legally binding instrument. Conventions that are legally binding in Canada such as the Convention Against all forms of Racial Discrimination or the Convention Against Torture for example, refer only generally and vaguely to the obligations on states to provide remedies for mass violations of the human rights they contain.

Domestic law in Canada is equally vague. The Canadian Charter of Rights and Freedoms[13] stipulates that 'Anyone whose rights or freedoms … have been infringed or denied may … obtain such remedy as the court considers appropriate and just in the circumstances.[14] The remedies available under the Charter include invalidating laws, declarations, damage awards and restitution but to date there has never been a comprehensive set of remedies responding to mass Charter violations such as those achieved under the IRSSA. It is only in rare cases the court has ordered the government to pay individual compensation.[15]

The Canadian Human Rights Act[16] provides for financial compensation for individuals, non-financial measures and measures to address the broader public interest meant to have an impact on the broader community, and to prevent similar discrimination from happening in the future.[17] However, these statutory remedies are limited to discrimination in employment, accomodation and the provision of services and not meant to apply to large groups of people with multiple needs who have experienced systemic discrimination and genocidal crimes. Moreover, the human rights regime in Canada is based on an individual compliant model. If an individual is successful in proving a breach of the legislation the tribunal can compensate for the loss of the right to be free from discrimination, including the insult to dignity, feelings and self-respect. Compensation is also available for any special financial costs caused by the discrimination with the intention of putting a claimant back in the financial position he or she would have been in if the discrimination had not occurred. With its narrow range of protections and its predominantly individual focus, the Canadian Human Rights Act is not suitable for mass claims of human rights violations of the sort that were experienced by residential school survivors.

Domestic criminal law cannot provide remedies for victims of mass human rights abuse either. This is because the purpose of criminal law is to punish offenders and protect citizens from crimes against the state. The role of victims of crime in the criminal justice process is as witnesses to the crime only. Not being a victim-centred process, the criminal provides very little if any satisfaction other than official recognition that a crime was committed and some retribution through the punishment ordered by the court.

The common law of torts unlike criminal law, is meant to provide remedies for those injured by others through wrongful acts. Class action law suits have made tort the favoured area of law to seek reparations for mass human rights violations because class actions are a form of lawsuit in which one or more persons bring a claim to court in the hopes of representing others who have suffered a similar harm at the hands of the same perpetrator. The primary advantage of a class action law suit is that it is often the only realistic option to bringing justice to a cause. It allows a group of people to have access to the courts in situations where the case would be too expensive or too complex for one person to sue on his or her own.

Class actions also save court costs by only having one trial; and they deter wrong doing by making defendants pay large sums for any harm they may have caused to individuals in the class. The remedial principle underlying a class action lawsuit is the same as in an individual tort suit, namely, that wrongs causing injury give the victim the right to be placed in the positon they would have been in but for the wrong.

Many survivors of the Indian residential schools were represented by class actions for their physical, psychological and sexual abuse and the harms they caused. The heavy burden of proving causation in court however, even within a class action, falls on the individual representing the class. Proving causation on the balance of probabilities forty of fifty years after the event is difficult if not impossible to do. Courts are also constrained by limitations periods, crown immunity, charitable immunity and vicarious liability constraints that limit the claimant's ability to succeed. In Canada, recent adjustments to the law of time limits on abuse[18] have made it easier for individual claimants and class action claimants to succeed in historic abuse cases. Similarly, the loosening of crown[19] and charitable immunity[20] from the law has helped to get the defendants before the courts to answer to abuses. Vicarious liability too, has undergone judicial reform which broadened the liability of employers for the tortious acts of their employees,[21] yet the principle is still limited by the requirement to show the abuser's employment has s 'strong connection' to the facilitation of the abuse.[22] Joint and severable liability legislation making co-defendants liable for the full amount of tort claims is very helpful to class actions claimants especially in cases like the residential school case where there were multiple defendants.

However, even with these improvements to the common law, court processes are still very problematic for victims of mass abuses. Many of the IRS victims fell outside of tort parameters not able to claim remedies for the harms they said were the most egregious, namely the profound loss of their family life, and loss of language and culture. The remedies they wanted, such as a common experience payment, advance payments for the ill or elderly, truth telling, healing, commemoration, memorialization, apologies or intergenerational harms are unattainable under the common law of torts. The only reparations that are available are compensation awards for physical, sexual and psychological abuse.

The limited reach of tort law with respect to the actual harms caused by the mass human rights abuses is also very troubling. In the residential school context, profound systemic harms designed to destroy the existence of indigenous identity through cultural genocide[23] of survivors and their descendants could not be addressed in a court of law. The sharp divergence between generally accepted understandings of moral wrongdoing and legal rules are most obvious where mass violations of human rights and harms such as those experienced in the residential school tragedy are in issue.

Despite the legal hurdles and limitations, however, a flood of litigation launched by former students of residential schools began in 1990 shortly after Phil Fontaine, Grand Chief of the Assembly of Manitoba Chiefs, spoke out on national television about the abuses he and thousands of other indigenous children endured at the 150 residential schools across Canada.[24] This was the first time any indigenous leader exposed the truth of residential schools, calling for an inquiry and an opportunity for survivors to relate their experiences to the Canadian public. The result was thousands of people coming forward, both individually and in class actions, to file tort actions in the courts. In Alberta, gridlock ensued when 1,479 actions involving 4,000 plaintiffs were filed, prompting one judge to comment that it would take 53 years of litigation to clear the dockets.[25]

In the midst of this flood, the Court of Appeal of Ontario certified a class action[26] for one residential school, indicating a distinct possibility that a general liability could follow across the country.

The Assembly of First Nations (AFN) under the leadership of Chief Phil Fontaine who had become the National Chief in 1997, then entered into the fray filing a massive class action in the courts on behalf of survivors while opening up channels of discussion at the political level.[27] It was at this point that the opportunity arose to create a model of reparations that would take into account indigenous legal principles, restorative justice principles, feminist legal principles in addition to conventional legal processes, to create a model and a range of reparations that could not only reflect the international guidelines, they could address the specific and diverse losses of the survivors and their families and communities.

Settlement was a viable option because the federal government, concerned about the gridlock in the courts and its uncertain liability on the one hand, and the claimants, concerned about delay, cost of litigations, aging and impoverished survivors and high risk litigation on the other, had good reason to consider it.

It is interesting to note that it was not until the AFN filed their class action that indigenous remedies and values were discussed by any of the parties to the litigation. The mainstream class action and independent lawyers were focused exclusively on the tort model. There was no recognition in their pleadings or their settlement proposals that there were other remedies survivors desperately wanted. The truth commission negotiations, for example, had no participation by any plaintiff lawyers other than the AFN team. Similarly, negotiations for commemoration, healing, memorialization and apologies were negotiated solely by the AFN on the plaintiff's side even though there were more than 80 lawyers representing various class and individual actions in the room.

In addition to demanding reparative indigenous remedies in the negotiations, the AFN introduced indigenous processes into the negotiations. Where class action lawyers decided among themselves what the best legal and remedial strategies should be, the AFN reached out to thousands of survivors, elders, community members and intergenerational survivors from coast to coast to coast to ascertain what they wanted from the process and how they wanted the process to be conducted.[28] The instructions the AFN received were to conduct itself consistent with a comprehensive set of values.

After the consultations with the indigenous community members it was apparent that all five components of reparations set out by the UN principles – restitution, compensation, rehabilitation, satisfaction, and guarantees of non-repetition – were required in any settlement agreement, as well as specific indigenous processes and remedies such as community designed healing projects, commemorative events, ceremonies, memorials, and community truth telling sessions among themselves.

The AFN also ensured that their negotiating team was diverse, made up of the National Chief, survivors and intergenerational survivors, an elder advisor, academics with international human rights expertise, and some lawyers. This was unlike the other legal teams who were comprised almost exclusively of white, male lawyers from large urban law firms.

This IRRSA experience demonstrates how difficult it is for lawyers, even in the face of dealing with indigenous peoples as their clients, to challenge western legal conceptions of what wrongs or injustice consist of and to think outside of their own experience. As Professor Carrie Menkle-Meadow points out, old methods of lawsuits, trials, and affixing blame are clearly not enough to repair the harms of past injustices to indigenous peoples.[29] With the expansion of standing requirements in modern human rights law,[30] indigenous groups around the world are now more than ever before, taking cases forward on behalf of themselves, their peoples, and their land in international, national, regional, local, and even private and quasi-private tribunals for claims involving property and other cultural and resource expropriation.[31] This is why the Indian Residential School Settlement Agreement,[32] stands as a very important example of how indigenous claims can and should be remedied in the future.

The judicial approval of the IRRSA claims in the Court ordered settlement agreement recognizing claims for loss of language, loss of culture and intergenerational harms, as well as the remedies of a common experience payment, truth commission, healing funds, commemoration, research centre, apologies and the like[33] raises important jurisprudential questions about what new causes of actions and remedies claims can and should be recognized by judges and courts and how future claims should be designed. It also raises questions about the role of the legal profession, law and legal processes in the pursuit of justice, reconciliation and restoration of peoples after mass harms has been done. The impacts of group harm, such as those harms to the indigenous peoples of Canada caused by centuries of colonialism, requires a re-thinking of fundamental conceptions of individualism, justice and what wrongs and injustice consist of.

Coming to terms with the limitations of the traditional forms of law and legal remedies is upon us, especially as indigenous peoples increasingly assert their rights and move toward nation to nation relationships which include demands to recognize indigenous legal traditions.[34] In Canada as well as in other colonizing countries, the need to resolve historic indigenous claims is an increasing imperative.

In conclusion, until the law and legal process are changed, any attempt to address mass human rights violations against indigenous peoples should be resolved by way of a settlement agreements or by legislation.[35] Tort remedies on their own are insufficient to meet the needs of the survivors and must be linked to other remedies, such as truth-seeking, institutional reform, accountability mechanisms, commemoration, healing and memorialization as was done in the IRSSA. Otherwise, simply applying tort law and providing damages will probably be counter-productive and even re-victimizing. For example, if tort damages are paid for certain harms but there is no opportunity for the victims to tell their stories and have them recorded, the payments may be perceived as an effort to buy victims' silence. This could allow perpetrators to deny their wrong doing, commit similar abuses in the future and make a mockery of the initiative to repair the wrong. Likewise, a reparations programme that fails to ensure that perpetrators are held accountable effectively asks victims to trade away their right to justice in order to receive an amount of money.

The Indian Residential School Settlement Agreement stands as an example of what can be accomplished when authentic voices of the victims of mass human rights abuses are heard and respected. The Agreement is a combination of compensation, restitution, satisfaction, rehabilitation and guarantees of non-repetition combined with indigenous legal traditions and values.

The common experience payment compensated for loss of language and culture and loss of family life. The physical, sexual and psychological abuses were generously compensated through an independent adjudicated process tailored to address survivor needs. Satisfaction was addressed by the Truth and Reconciliation Commission and the creation of the research centre at the University of Manitoba. The 95 Calls to Action[36] addressed the need for restitution and reform of all major Canadian institutions. Rehabilitation was addressed through financial support for the Aboriginal Healing Foundation and the education credits available to intergenerational survivors. Guarantees of non-repetition were addressed through individual apologies as well as through the historic apology in the House of Commons in the presence of all the indigenous national leaders and the apology by the Pope of the Roman Catholic Church.[37]

Reparations properly done, have the potential to build trust and restore dignity. They also serve to provide a measure of justice directly to victims, offering them a future that may alleviate, to some extent, the suffering they have endured and provide some form of reconciliation. Without a direct focus on victims' needs along with clear acknowledgment and recognition of the wrongs committed and their impacts beyond individuals, laudable objectives of reconciliation and healing will probably fail. In this regard the IRSSA, although not perfect, should stand as an example for the future.

Notes

* FRSC, QC. The author was the chief negotiator for the Assembly of First Nations for the historic Indian Residential School Settlement Agreement. She was the principle architect of the Truth and Reconciliation Commission in this role.

1 Department of the Interior, Canada, Annual Report, 1880; 'Report on Industrial Schools for Indians and Half-Breeds', Nicholas Flood Davin, 14 March 1879. See also Truth and Reconciliation Committee (TRC) Final Report vol 1, pp. 56–57, www.myrobust.com/websites/trcinstitution/File/Reports/Volume_1_History_Part_1_English_Web.pdf.

2 Library and Archives Canada, RG 10, vol 6810, file 470-2-3, vol 7, Evidence of D. C. Scott to the Special Committee of the House of Commons Investigating the Indian Act amendments of 1920 (L-2) (N3). Also cited in TRC Final Report, vol 1 at p. 3.

3 The term has been attributed to Duncan Campbell Scott but more accurately it originated in the US military. See TRC Report, supra.

4 If success can be measured by the numbers of people who opted into the agreement, 98% of the survivors made this choice rather than proceeding through the courts; 77% of 105,530 applicants for the common experience payment received payment; more than 59% of the 38,099 individual assessment process claims were successful with the average payout being $111,758.00. The TRC activities included 155,000 visits to national events; regional events held 238 days of local hearings in 77 communities across Canada. The Commission received over 6,750 statements from survivors and their families. See www.aadnc-aandc.gc.ca/eng/1315320539682/1315320692192 for more statistics.

5 See Basic Principles and Guidelines on the Right to a Remedy and Reparation for Victims of Gross Violations of International Human Rights Law and Serious Violations of International Humanitarian Law, adopted by General Assembly

Resolution 60/147 of 16 December 2005, www2.ohchr.org/english/law/ remedy.htm. See also the International Covenant on Civil and Political Rights, art. 2(3), and the Convention Against Torture and Other Cruel, Inhumane, or Degrading Treatment or Punishment, art. 14(1), on state responsibility to provide redress and reparations to victims.

6 See http://legal.un.org/avl/ha/ga_60-147/ga_60-147.html.

7 Ibid.

8 Some of the harms listed as compensable in the compensation model are Loss of self- esteem, pregnancy, forced abortions, forced adoptions, psychotic disorganization, PTSD, self- injury, sexual dysfunction, inability to form or retain relationships, eating disorders, severe anxiety, guilt or self-blame, lack of trust in others, addictions, nightmares, aggression, hypervigilance, anger, retaliatory rage, and humiliation.

9 The Settlement Agreement provides for loss of opportunity described as one of chronic inability to obtain employment; chronic inability to retain employment, periodic inability to obtain or retain employment, inability to undertake or complete education or training resulting in underemployment and/or unemployment, or diminished work capacity. Claims can alternatively be made for actual income loss. For the settlement agreement, see www.residentialschoolsettlement.ca/english_index.html.

10 Tort law uses the legal device of foreseeability to limit liability to only those the wrongdoer can foresee would be affected by the acts or omission in question. In most cases, injuries to third parties are considered to be too remote. *The Wagon Mound (no. 1) v. Miller S. S. Co. Pty. Ltd.* [1966] 2 All E.R.709, [1967] 1 A.C. 617.

11 TRC Final Report, supra, note 1 at. 90–100.

12 Supra, note 5.

13 Constitution Act, 1982 http://laws-lois.justice.gc.ca/eng/Const/page-15.html.

14 Canadian Charter, ibid, section 24.

15 See Koster and Pyzer, Remedies Pursuant to section 24 and 25 of the Canadian Charter of Rights and Freedoms, www.torontodefencelawyers.com/blog/general-category/remedies-pursuant-to-the-canadian-charter-of-rights-and-freedoms-ss-24-and-52.

16 R.S.C., 1985 c.H-6.

17 Canadian Human Rights Commission, Your Guide to Understanding the Canadian Human Rights Act, www.chrc-ccdp.gc.ca/eng/content/your-guide-understanding-canadian-human-rights-act-page1.

18 *M(K) v. M(H)* [1992] 3 SCR 6.

19 Crown Liability and Proceedings Act, RSC 1985, c C-50 s1; 1990, c.8 s. 21(3) (b) (i). Crown immunity still exists for intentional torts committed prior to 1949 and for policy decision as opposed to operational ones. See *Just v. BC* [1989] 2 SCR 1228.

20 *Blackwater v. Plint*, 2005 SCC 58 at para. 61; Re Winding Up of the Christian Brothers of Ireland in Canada (2000) SCCA No.277 (QL).

21 *Bazley v. Curry* [1999] 2 SCR 534.

22 *Jacobi v. Griffiths* [1999] 2 SCR 570.

23 Genocide is an attempt to destroy a people, in whole or part – is a crime under international law. The United Nations' Convention on the Prevention and Punishment of the Crime of Genocide, adopted in 1948. It does not use the phrase 'cultural genocide', but says genocide may include causing serious mental harm to a group. Supreme Court Chief Justice Beverley McLachlin used the term when she said Canada attempted to commit 'cultural genocide' against aboriginal peoples in a speech 28 May 2015. Former Liberal prime minister Paul Martin used the term in 2013 describing residential schools for aboriginal children when he testified before the Truth and Reconciliation Commission.

24 Phil Fontaine's Shocking Testimony of Sexual Abuse, CBC Archives, www.cbc.ca/archives/entry/phil-fontaines-shocking-testimony-of-sexual-abuse.

25 Terrance McMahon, Remarks, University of Toronto conference, 18 January 2013, cited by Mayo Moran in the Role of Reparative Justice in Responding to the Legacy of Indian Residential Schools, (2014) 64 *UTLJ* 529.
26 *Cloud v. Canada* (2004) 73 OR 93rd0 401.
27 Several class actions had been filed in the courts but none had indigenous survivors on their negotiating teams.
28 G. Sigurdson, Reconciliation and Healing: Alternative Resolution Strategies, www.glennsigurdson.com/wp-content/uploads/2016/06/Reconciliation_healing2.pdf.
29 Carrie Menkle-Meadow, Unsettling The Lawyers: Other Forms of Justice in Indigenous Claims of Expropriation, Abuse and Injustice (2014) 64 *UTLJ* 620.
30 See for example, Christopher Whytock, Some Cautionary Notes on the Chevronization of Transnational Litigation (2013) 1 *Stanford LJ of Complex Litigation* 467.
31 Supra, note 29.
32 See www.residentialschoolsettlement.ca/english_index.html.
33 Ibid., Court Judgments.
34 See for example, Lisa Chartrand, Accommodating Indigenous Legal Traditions, Indigenous Bar Association, 31 March 2005 www.indigenousbar.ca/pdf/Indigenous%20Legal%20Traditions.pdf.
35 The Indian Residential School Settlement Agreement was a court ordered settlement agreement. This was the preferred options because of the lack of trust of government's ability to draft and implement appropriate legislation.
36 Truth and Reconciliation Commission, 95 Calls to Action, www.trc.ca/websites/trcinstitution/File/2015/Findings/Calls_to_Action_English2.pdf.
37 The apologies became part of the overall settlement negotiations between the AFN, the government and the Roman Catholic church.

7 The struggle over the Dakota Access Pipeline in the context of Native American history

Joseph W. Dellapenna

On 23 June 2014, Energy Transfer Partners through its subsidiary Dakota Access LLC announced a plan to construct the 'Dakota Access Pipeline', from the Bakken oil shale fields in North Dakota across South Dakota and Iowa to an oil tank farm in Illinois. When construction began in May 2016, there were already numerous lawsuits pending against the project as well as further regulatory hurdles. A group of 200 from the Standing Rock Sioux reservation made a dramatic ride from their reservation to what would become the protest campsite on 1 April 2016 (Woolf 2016). Resistance to the pipeline failed in the end, although some court challenges persist. The struggle over the Dakota Access Pipeline is not an isolated event, however, but merely the latest chapter in an ongoing struggle between Native Americans and people of European extraction that began in 1492 and will continue into the future. This chapter examines the dispute between the Native Americans across whose sacred site the pipeline runs in the context of the more than five centuries of interaction between Native Americans and Europeans or European Americans, and in particular the legal theories that developed in US courts on Native American relations.

The Europeans come to America

When Christopher Columbus made landfall in the Caribbean area, he found a hemisphere peopled by approximately 50,000,000 persons, with some estimates ranging as high as 112,000,000 (Deneven 1992). These people were divided into numerous groups, generally called tribes, speaking one of hundreds of languages (Campbell 2000). While we can today reconstruct in general terms were each was located in 1492, the reconstruction cannot be precise. A tribe's lands often overlapped with the other tribes' lands, particularly when they lived as hunter–gatherers. Relation were sometimes friendly and sometimes hostile, and many tribes were moving to new living areas, often through bloody clashes with the prior inhabitants of the areas (Anderson 1980).

The tribes had highly varied cultures (Treuer et al. 2010). Some had advanced agricultural techniques supporting dense urban populations with large stone temples, palaces, and other buildings, ranging from central Mexico to the Andean region of South America (Dean 2010; Phillips 2015). The influence of these cultures reached into the what is now the United States. Farming communities living in stucco apartment buildings were found in the southwest of the United States, while in the southeast agriculture supported large settlements featuring large earthen mounds and thatched buildings. In the northeast of the United States and eastern Canada, the people were mostly hunter–gatherers with small-scale farming, living in crude shelters; the more advance Iroquois had culture that resembled that of the southeast of the United States. In the northwest, there were dense, forest-dwelling communities that lived in log houses and ate salmon, whales, and acorns. In parts of California and across the plains, the people were hunter–gatherers scattered in small numbers in less hospitable lands. How these various cultures and their peoples would have evolved absent the European impact is speculative.

The Europeans caused massive disruption throughout the Americas, some intentionally and some not. Within a century of the arrival of Europeans, the Native American population reduced drastically, in some places by as much as 90 percent (Newson 1993), often as the result of conquests in which large numbers died from slaughter, slavery, starvation (de las Casas 1642). Even more devastating were European diseases for which the Native Americans had no immunities. Small pox was a particular scourge, traveling far ahead of the Europeans, depopulating large areas before the first attempts of Europeans to subdue or settle these places (Daniels 1992).

Hernando de Soto led the first European effort to conquer or settle in the current United States (Hudson 1998). De Soto landed in Florida in 1539 and for three years tried to conquer the southeast, leaving disease behind. Sometime around 1615, fishermen from Europe introduced the pox into the northeast of the United States, which enabled the Pilgrims in 1620 to survive on the food stocks of villages left empty by the epidemic (Marr and Cathey 2010). Not all of the impacts of the Europeans were bad. The escape of horses in New Mexico during the Pueblo revolt of 1680–1692 (Knaul 1995) allowed development of a horse culture across the plains and the movement of native peoples from the forests into the plains (Haines 1988). Many today think of that horse culture as the quintessence of Native American culture but it lasted little more than a century before European contact destroyed it. The Lakota were one of peoples who moved onto the plains during this time (Anderson 1980).

British colonial policy towards the Indians

In 1492 there were 10 to 12 million people living in what is now Canada and the United States (some estimates are higher) (Daniels 1992). These numbers declined rapidly after the arrival of the Europeans. The British and the French, who seized and settled the eastern half of North America between 1600 and 1750, applied the 'Doctrine of Discovery' to justify sovereignty over the newly 'discovered' lands. By that doctrine, the act of discovering lands held by non-Christians vested title in the Christian sovereign whom the 'discoverer' represented (Miller and Ruru 2012). Along with sovereignty, private ownership of the land vested in the sovereign subject to a right of occupancy in the original inhabitants that lasted until terminated by the new sovereign (*Johnson v. M'Intosh* 1823, p. 585).

The British and the French warred against each other, with tribes allying themselves with the European powers to fight increasingly unwanted settlers (Anderson 2001). After the British ousted the French from west of the Allegheny Mountains (Dowd 2004), the British sought to stabilize their relations with the native tribes by excluding colonial governments from dealing with the tribes and by closing 'Indian country' (land west of the Allegheny Mountains) to European settlement (Fenge and Aldridge 2015). The colonists and colonial governments had some say only within the short watersheds that drained directly into the Atlantic. The attempt to close the frontier became a major grievance leading to the American Revolution (Schmidt 2014).

The legal status of the tribes in the United States

The US Constitution adopted in 1787 conferred on the federal government many of the powers claimed by London before the American Revolution, to the consternation of 'anti-Federalists' who opposed ratification of the Constitution (Edling 2003). Among those powers was the authority '[t]o regulate Commerce with foreign Nations, and among the several States, and with the Indian Tribes' (US Constitution, Art. I, §8, cl. 3). This does not explicitly make federal authority exclusive. The US Supreme Court concluded that it was exclusive as against the states, reasoning that because the federal authority over commerce with foreign nations must necessarily be exclusive, that must hold true for relations with Indians as well because the authority is granted in a single clause and without distinction among the forms of commerce (*Gibbons v. Ogden* 1824, p. 90). That conclusion holds today (*Michigan v. Bay Mills Indian Community* 2014, p. 2030). That decision is binding on all courts and other forms of government in the United States, state as well as federal, because the US Constitution and 'laws made pursuant thereto' are the 'supreme law of the land', overriding any state or local law to the contrary (US Const. art. VI, §2).

Even after1824, state and local governments continued to encroach on tribal lands (*City of Sherrill v. Oneida Indian Nation* 2005). Private persons of European descent also seized Indian lands, sometimes cooperating with state and local governments and sometimes defying them (Ostler 2011). In an effort to minimize conflict, the federal government undertook a policy of restricting tribes to reservations – generally on less desirable lands (Petrikowski and Edwards 2016). The southeast of the United States was particularly ripe for such conflict (Bunn and Williams 2008). European American settlers were growing in numbers and wealth, but they confronted 'the five civilize tribes', remnants of the mound-builder cultures that existed before de Soto's expedition (Foreman 1982). The five civilized tribes (the Cherokees, Chickasaws, Choctaws, Creeks, and Seminoles) developed farms and plantations, owned slaves, and built large and growing towns (Krauthamer 2015). Sequoyah, a Cherokee, invented an alphabet which is still taught in tribal schools and in which newspapers and books still are published.

In its first decision specifically addressing title to lands inhabited by Native Americans, the US Supreme Court held that in order to preserve the peace between them a conveyance from an 'Indian' to a white person could not be valid (*Johnson v. M'Intosh* 1823).This decision made clear that for a white settler to get good title to land taken (even by purchase) from a Native American, the Indians would have to be completely (and, if necessary, forcibly) removed from the land. The state of Georgia responded by enacting a statute in 1830 to invalidate all laws enacted by, and land titles held by, Native Americans in Georgia. European American began seizing Native American properties in Georgia, killing any resisting Native Americans (Williams 1994). The Cherokees were too civilized (or too aware of the futility of armed resistance) to go to war. Instead they went to court, giving the US Supreme Court the opportunity to define the legal status of Native American tribes under American law.

The tribes, the Court decided, are 'domestic dependent nations' with rights of occupancy but not title to their lands (*Cherokee Nation v. Georgia* 1831, p.13; *Worcester v. Georgia* 1832, pp. 552, 561). Tribes were declared wards under the guardianship of the US who could not sell or even lease their lands without the approval of the US Government. (*United States v. Mitchell* 1983). Too often, wardship involved mistreatment rather than the protection that a ward would expect (Rey-Bear and Fletcher 2017). This relationship did not extinguish tribal sovereignty (O'Connor 1997), and under the Constitution, state and local governments had no authority over the tribes. The Cherokee won their lawsuits but that did not prevent President Andrew Jackson from expelling them and the other civilized tribes from their landholdings and their forced migration to Oklahoma on what the Cherokee remember as 'The Trail of Tears' (Ehle 1997).

Well into the nineteenth century, relations between the United States and the tribes were governed by treaties. The federal government ended the practice of 'negotiating' treaties in 1871, and relations thereafter were governed by statutes enacted by Congress (Act of 3 March 1871). Yet the idea that Native Americans formed dependent nations that were not strictly a part of the United States continues to this day. One consequence is that Indians were not counted for representation in Congress, were not citizens of the United States, and were not allowed to vote. (*Elk v. Wilkins*, 1884). This only changed when Indians became citizens by act of Congress in 1924 (Act of 2 June 1924).

US–tribal legal relations in the late twentieth and early twenty-first centuries

After 1871, federal policy (through statutes) vacillated between assimilation and restoration of tribal autonomy (Hoxie 1984; Kelly 1983). This vacillation helped mask mismanaged funds and property, a serious problem that is not yet fully resolved (Rey-Bear and Fletcher 2017). In the later years of the twentieth century, three other major areas of legal dispute not hitherto considered arose between the tribes and the United States: (1) protection of sacred or religious sites; (2) protection of fishing and hunting rights; and (3) casinos.

1. Many Native American religious sites are located on land owned by the federal government. While the Constitution's free exercise of religion guarantee (US Const. Amend. 1) applies to the tribes and their members, the federal government generally is free to manage its property as it chooses (*Lyng v. Northwest Cemetery Protective Ass'n* 1990). Furthermore, laws of general applicability can prohibit a religious practice, so long as there is no discrimination against a religion or religion generally (*Employment Division v. Smith* 1990). Congress reacted to this last decision by enacting a statute that bars the imposition of a 'substantial burden' on religious practices unless it is justified by a 'compelling state interest" and the government uses the 'least restrictive means' to accomplish its purpose (*Religious Freedom Restoration Act* 1993, §2000cc(a)(1)).
2. Guarantees of hunting, fishing, and gathering rights, even on private land, were included in many treaties between the federal government and the tribes. Such rights continue after the area becomes part of a state and even after dissolution of a reservation (*Minnesota v. Mille Lacs Band* 1999). Yet these treaty rights do not altogether preclude state regulation their hunting and fishing (*Puyallup Tribe, Inc. v. Washington Dep't of Game* 1977). State regulation is allowed if it is for conservation and if the Indians are given an appropriate share of the allowable catch. The problem with such vague standards is how to ensure that the Native American tribes have a voice in these decisions. The tribes found a way through litigation (*Mille Lacs*; *Puyallup*). Courts enforce the tribal rights over the objections of private landowners. A current dispute is over the Trump administration's plan to reduce the size of several National Monuments, created to protect scientific, historic, or religious sites (Friedman 2017). Secretary of the Interior Ryan Zinke insists that the concerned Native American tribes have been consulted and are 'happy' with the proposal, although Ethel Branch, the Attorney-General for the Navajo Nation, expresses the opposite opinion on both counts (Capriccioso 2017). The dispute captures the fraught federal-tribal relationship – the tribes are subordinate to the federal government which speaks 'on their behalf' even when the tribes do not agree with what the federal government says (Siegal 2017).
3. In 1988, Congress authorized Native American tribes to open casinos (*Indian Gaming Act* 1988). These rapidly proliferating casinos soon became a primary source of revenue for the tribes (Green 2017). The act of Congress in question, however, allows some regulation by states, which has opened another source of litigation over tribal rights (*Michigan v. Bay Mills Indian Community* 2014).

The legal dispute over the Dakota Access Pipeline

When the Lakota first encountered large-scale white encroachment on their lands, they accepted a large reservation covering half South Dakota and large areas of Nebraska, North Dakota, and Wyoming (Treaty of Laramie 1851). A later treaty reduced the reservation while guaranteeing hunting and fishing rights (Treaty of Laramie 1868). Further reductions and divisions left the Lakota with (among other reservations) the Standing Rock Reservation that straddles the North and South Dakota border (Lazarus 1991). The steady diminution of the reservation lands did not preclude certain claims by the tribes to rights outside the reservation such as the right to block actions that would impair the ecological integrity or sustainability of the reservations, and that would interfere with sacred sites. Thus, most of the foregoing strands of legal thought come together in the dispute over the Dakota Access Pipeline, along with general regulatory authority over interstate pipelines. We begin with the general regulatory authority over the pipeline.

There is no comprehensive federal regulation of oil pipelines. Energy Transfer Partners sought to avoid creating an opening for federal regulation by placing 99 per cent of the pipeline on private lands (*Standing Rock Sioux Tribe v. US Army Corps of Engineers* 2016, p. 7). It was unable to avoid two federal connections. First, it had to use some federal lands, bringing into play the *Mineral Leasing Act of 1920*. Second, the pipeline would cross navigable waters of the United States, bringing into play the *Rivers and Harbors Act of 1899*. Decisions under these acts in turn trigger the *National Environmental Policy Act of 1969* and the *National Historic Preservation Act of 1966*. The *Rivers and Harbors Act* requires a permit from the US Army Corps of Engineers, but the Corps had already issued a 'nationwide permit' for projects that involve a half-acre (0.2 hectare) or less, subject to consultation with affected Native American tribes if a project would affect tribal treaty rights or cultural resources (US Army Corps of Engineers 2012).

Energy Transfer Partners presented a permit application to the Corps that purported to analyze the legal issues and provided an environmental impact assessment. The Corps accepted the analysis and applied its 'nationwide permit' (*Standing Rock Sioux Tribe v. US Army Corps of Engineers* 2016, p. 7). The Sioux of the Standing Rock Reservation challenged the issuance of the permit, but their challenge failed. (*Standing Rock Sioux Tribe v. US Army Corps of Engineers* 2016). The resulting protests involved the creation of a permanent campsite astride the proposed route (the protest campsite was off reservation lands; Healy 2016). The protests drew in other Native Americans and even a contingent of US military veterans (Archambault 2016). The Obama administration then declined to grant a lease for the right-of-way across the federal lands under the *Mineral Leasing Act* and Assistant Secretary of the Army Jo-Ellen Darcy ordered the Corps to reconsider its permit because of harmful effects on 'treaty resources', and ordered a full environmental impact assessment (Healy and Fandos 2016).

On 20 January 2017, Donald Trump became president of the United States. On 24 January 2017, the new administration directed issuance of the permit and the lease forthwith (Baker and Davenport 2017). The Corps issued a 'Finding of No Significant [environmental] Impact' on 26 January 2017 and issued the permit on 7 February 2017 (Turkewitz 2017). Test flows of oil in the pipeline began by 28 March 2017 (Donovan 2017), with commercial flows following on 1 June 2017 (Slattery 2017) The pipeline had already experienced its first major spill on 10 May 2017, however, with more to follow (Levin 2017).

Conclusion

Local authorities cleared the protest campsite on 23 February 2017 (Iaconangelo 2017). The protesters had been called 'criminals' and even 'terrorists' (Hardy 2017), but when the government sought to prosecute the protestors for trespass or other crimes, most protesters were acquitted (Grueskin 2017). A federal judge ruled on 14 June 2017 that the permit was illegal without a proper environmental impact statement, but the judge declined to enjoin the operation of the pipeline (*Standing Rock Reservation Tribe v. US Army Corps of Engineers* 2017). The dispute over the impact statement continues. The end to the confrontation perhaps was when David Archambault, the elected chairman of the Standing Rock Reservation who had led the resistance, lost his bid for reelection (Holdman 2017). Even if that perhaps marks the end of the current confrontation, there will no doubt be others when the dominant culture once again disrupts the lives and livelihoods of Native Americans. The only questions are where, when, and to what purpose.

References

Legal materials

Act of 2 June 1924. *US Code tit.* 8, §1401(b).
Act of 3 March 1871. *US Code* tit. 25, §71.
Bear Lodge Multiple Use Ass'n v. Babbitt 1999 (10th Circuit Court of Appeals). *Federal Reporter, 3d series.* 175:814–822.
Cherokee Nation v. Georgia 1831. *US Reports* 30 (5 Pet.):1–54.
City of Sherrill, New York v. Oneida Indian Nation 2005. *US Reports* 544:197–227.
Elk v. Wilkens 1884. *US Reports* 112:94–123.
Employment Division v. Smith 1990. *US Reports* 494:872–921.
Gibbons v. Ogden 1824. *US Reports* 22(9 Wheat.):1–94.
Indian Gaming Act 1988. *US Code* tit. 25, §§2701–2713.
Johnson v. M'Intosh 1823. *US Reports* 21 (8 Wheat.) 543–605.
Lyng v. Northwest Indian Cemetery Ass'n 1988. *US Reports* 485:439–477.
Michigan v. Bay Mills Indian Community 2014. *Supreme Court Reporter* 134:2024–2056.
Mineral Leasing Act of 1920. *US Code* tit. 30, §§181–196.
Minnesota v. Mille Lacs Band 1999. *US Reports* 526:172–226.
National Environmental Policy Act of 1969. *US Code* tit. 42, §§4321–4335.
National Historic Preservation Act of 1966. *US Code* tit. 16 §§470–470w-6.
Puyallup Tribe, Inc. v. Washington Dep't of Game 1977. *US Reports* 443:165–185.

Religious Freedom Restoration Act 1993. *US Code tit. 42*, §§2000bb-1–2000cc-5.
Rivers and Harbors Act of 1899. *US Code* tit. 33, §§403–430.
Standing Rock Sioux Tribe v. US Army Corps of Engineers 2016. *Federal Supplement, 3d series*. 205:4–37.
Standing Rock Sioux Tribe v. US Army Corps of Engineers 2017. *Federal Supplement, 3d series*. 255:101–122.
Treaty of Laramie 1851. *Statutes at Large* 11:749.
Treaty of Laramie 1868. *Statutes at Large* 15:635.
US Army Corps of Engineers 2012. *Decision Document: Nationwide Permit 12* (13 February 2012),www.usace.army.mil/Portals/2/docs/civilworks/nwp/2012/NWP_12_2012.pdf13), www.usace.army.mil/Portals/2/docs/civilworks/nwp/2012/NWP_12_2012.pdf.
United States v. Jicarilla Apache Nation 2006. *US Reports* 564:162–209.
United States v. Mitchell 1983. *US Reports* 463:206–238.
Worcester v. Georgia 1832. *US Reports* 31 (6 Pet.):515–597.

Other Sources

Anderson, Fred 2001. *The Crucible of War: The Seven Years War and the Fate of Empire in British North America, 1754–1766* (reprint edn). New York: Vintage Books.
Anderson, Gary Clayton 1980. 'Early Dakota Migration and Intertribal War: Revision', *Western Historical Quarterly* 11(1):17–36.
Archambault, David II 2016. 'Taking a Stand at Standing Rock', *New York Times* (25 August), p. A19.
Baker, Peter and Coral Davenport 2017. 'Trump Revives Keystone Pipeline Rejected by Obama', *New York Times* (25 January), p. A1.
Bunn, Mike and Clay Williams 2008. *Battle for the Southern Frontier: The Creek War and the War of 1812*. Mt. Pleasant, SC: Arcadia Publishing.
Campbell, Lyle 2000. *American Native Languages: The Historical Linguistics of Native America*. Oxford: Oxford University Press.
Capriccioso, Rob 2017. 'Zinke Says Tribes Are 'Happy' to Have Bears Ears Modifications; Tribes Disagree', *Indian Country Today* (13 June), available at https://indiancountrymedianetwork.com/news/native-news/zinke-says-tribes-happy-bears-ears-modifications-tribes-disagree.
Daniels, John D. 1992. 'The Indian Population in North America', *William and Mary Quarterly* 49:298–320.
Dean, Carolyn J. 2010. *A Culture of Stone: Inka Perspectives on Rock*. Durham, NC: Duke University Press.
De las Casas, Bartolomé 1642. *The Devastation of the Indies: A Brief Account*. (Herma Brifffault trans. 1974). London: Continuum Books.
Deneven, William M. (ed.) 1992. *The Native Population of the Americas in 1492* (2nd edn). Madison, WI: University of Wisconsin Press.
Donovan, Laruen 2017. 'Oil Flowing under Oahe', *Bismarck (N.D.) Tribune* (28 March), p. A1.
Dowd, Gregory Evans 2004. *War under Heaven: Pontiac, the Indian Nations, and the British Empire*. Baltimore, MD: Johns Hopkins University Press.
Edling, Max M. 2003. *A Revolution in Favor of Government: Origins of the U.S. Constitution and the Making of the American State*. Oxford: Oxford University Press.
Ehle, John 1997. *The Trail of Tears: The Rise and Fall of the Cherokee Nation*. New York: Doubleday.

Fenge, Terry and Jim Aldridge 2015. *Keeping Promises: The Royal Proclamation of 1763, Aboriginal Rights, and Treaties of Canada.* Kingston, ON: McGill-Queen's University Press.

Foreman, Grant 1982. *The Five Civilized Tribes.* Norman, OK: University of Oklahoma Press.

Friedman, Lisa 2017. 'Trump Plans to Shrink Two National Monuments in Utah', *New York Times* (28 October), p. A10.

Green, Gary 2017. *Oceola's Revenge: The Phenomena of Indian Casinos.* New York: Brick Tower Press.

Grueskin, Caroline 2017. 'Tribal Leaders found Not Guilty on Protest Charges', *Bismarck Tribune* (1 June), p. A1.

Haines, Francis 1988. 'The Northward Spread of Horses among the Plains Indians', *American Anthropologist* 40(3):429–437.

Hardy, Kevin 2017. 'Are Pipeline Protesters Terrorists? Some Members of Congress Seek Charges', *Des Moines Register* (28 October), p. A11.

Healy, Jack 2016. 'North Dakota's Oil Pipeline Battle: Who's Fighting and Why?', *New York Times* (27 August), p. A12.

Healy, Jack and Nicholas Fandos 2016. 'Protestors Gain Victory in Fight over Dakota Access Pipeline', *New York Times* (5 December), p. A1.

Holdman, Jessica 2017. 'Archambault Loses Standing Rock Chairman's Seat', *Bismarck Tribune* (28 September), available at http://bismarcktribune.com/news/state-and-regional/archambault-loses-standing-rock-chairman-s-seat/article_346bcc71-353c-5560-b1c6-a87cc4e5af4f.html.

Hoxie, Frederick E. 1984. *A Final Promise: The Campaign to Assimilate the Indians, 1880–1920.* Lincoln, NE: University of Nebraska Press.

Hudson, Charles 1998. *Knights of Spain, Warriors of the Sun: Hernando de Soto and the South's Ancient Chiefdoms.* Athens, GA: University of Georgia Press.

Iaconangelo, David 2017. 'With Dakota Access Camp Cleared, Where Are Protesters Headed?', *Christian Science Monitor* (25 February).

Kelly, Lawrence C. 1983. *The Assault on Assimilation: John Collier and the Origins of Indian Policy Reform.* Albuquerque, NM: University of New Mexico Press.

Knaul, Andrew L. 1995. *The Pueblo Revolt of 1680.* Norman, OK: University of Oklahoma Press.

Krauthamer, Barbara 2015. *Black Slaves, Indian Masters: Slavery, Emancipation, and Citizenship in the Native American South.* Chapel Hill, NC: University of North Carolina Press.

Lazarus, Edward 1991. *Black Hills, White Justice: The Sioux Nation v. United States, 1775 to the Present.* New York: HarperCollins.

Levin, Saul 2017. 'Dakota Access Pipeline Has First Leak before Pipeline Is Fully Operational', *The Guardian* (10 May).

Marr, John S. and John T. Cathey 2010. 'New Hypothesis for Cause of Epidemic among Native Americans, New England, 1616–1619', *Emerging Infectious Diseases* 16(2):281–286.

Miller, Robert J. and Jacinta Ruru 2012. *Discovering Indigenous Lands: The Doctrine of Discovery in the English Colonies* (reprint edn). Oxford: Oxford University Press.

Newson, Linda A. 1993. *Proceedings of the British Academy* 81:247–288.

O'Connor, Sandra Day 1997. 'Lessons from the Third Sovereign: Indian Tribal Courts', *Tulsa Law Journal* 33:1–6.

Ostler, Jeffrey 2011. *The Lakotas and the Black Hills: The Struggle for Sacred Ground* (reprint edn). New York: Penguin Books.

Petrikowski, Nicki Peter and Judith Edwards 2016. *The Relocation of Native Peoples of North America*. New York: Enslow Publishing.

Phillips, Charles 2015. *The Complete Illustrated History of the Aztec and Maya: The Definitive Chronicle of the Ancient Peoples of Central America and Mexico – Including the Aztec, Maya, Olmec, Mixtec, Olmec, Toltec,and Zapotec* (3rd edn). London: Hermes House.

Rey-Bear, Daniel I.S.J. and Matthew L.M. Fletcher 2017. "We Need Protection from our Protectors': The Nature, Issues, and Future of the Federal Trust Responsibility to Indians', *Michigan Journal of Environmental and Administrative Law* 6:397–461.

Schmidt, Ethan A. 2014. *Native Americans in the American Revolution*. Santa Barbara, CA: Praeger.

Siegal, Josh 2017 'Navajo Nation Threatens to Sue Trump if he Shrinks Bears Ears National Monument', *Washington Examiner* (21 September), available at www.washingtonexaminer.com/navajo-nation-vows-to-sue-trump-if-he-shrinks-bears-ears-national-monument/article/2635214.

Slattery, Denis 2017. 'Oil to Flow, but Probe Looms, *New York Daily News* (1 June), p. 26.

Treuer, Anton 2010. *Indian Nations of North America*. Washington, DC: National Geographic.

Turkewitz, Julie 2017. 'Approves Construction of Dakota Access Pipeline', *New York Times* (8 February), p. A10.

Williams, David 1994. *The Georgia Gold Rush: Twenty-Niners, Cherokees, and Gold Fever*. Columbia, SC: University of South Caroline Press.

Woolf, Nicky 2016. 'Native American Tribes Mobilize against North Dakota Pipeline', *The Guardian* (2 April).

8 Navigating complexity, promoting health

Insights from the emergence of 'Ecohealth' and 'One Health'

Colin L. Soskolne, Martin J. Bunch, Colin D. Butler and Margot W. Parkes

An Overview of Ecohealth

As the combination of the words 'Eco' and 'Health' suggests, Ecohealth addresses the intimate interdependence of human health and the ecosystems in which people are embedded. This connection is now a commonplace observation in some areas of science and thought. In the recent past, however, there has been a growing disconnect between the two, so that today many humans, especially non-indigenous, still consider themselves as separate from the natural world in which they live.

In part, this disconnect is a symptom of the western tradition of reductionist science and mechanistic management that has evolved since the scientific revolution. This has trained us to reduce problems to their component parts, and to understand those parts in isolation. This, however, misses emergent properties that are evident only at the level of whole systems (Bavington 2002), such as many of the relationships that characterize the dependence of our health and well-being on the ecological contexts in which we are embedded. It is also exacerbated by factors such as technology and our built environment, to the point where some experience symptoms that may reflect disassociation from nature (Hofferth and Sandberg 2001; Louv 2005; Malone 2007). The importance of understanding the ecological context of both our homes and our health in ecological terms is increased when considering our urbanizing and technology-dependant world. These trends not only influence development and experiences at earlier ages, but have increased the demand for 'ecological literacy' in order to overcome the disconnect from the living systems on which we depend (Parkes and Horwitz 2016: 28).

On planet Earth, ecosystem services are the benefits that humans obtain from nature and that underpin their social and economic well-being, including breathable air, drinkable water and fertile soils to grow nutritious foods – all that sustains life (Millennium Ecosystem Assessment 2005). So much more are these connections recognized today that several countries have enshrined or begun to enshrine ecosystem protections in policies and also in constitutional law (Boyd 2012a, 2012b). Respect for these connections is necessary to protect the health and well-being of both present and future generations. Growing awareness of these connections has also facilitated an expansion of the public health discourse.

After more than a century of focus on diseases, their biologic causes, the correction of exposures (clean water and air) and facilitation of responses (immunizations and nutrition), there is as shift in the public health discourse towards a greater embrace of the concept of deeper determinants of health, an undercurrent which has long existed within social medicine (Butler 2013), and is now demanding increasing sophistication in active engagement across both social *and* ecological determinants of health (Hallström et al. 2015). Increased focus on the determinants of health reflects the longstanding need for attention to the more upstream causes of ill-health and premature death; they extend from the more proximate causes to the social, economic and environmental realms (McMichael 1999). These considerations have moved the discourse that frames public health science to an unprecedented level of complexity that, in many ways, evolved from the heightening of public concern about the environment that has grown since the 1960s.

The field of Ecohealth can be seen as a recent response to this challenge, seeking to reduce the separation between humans and ecosystems using a multifactorial, dynamic approach that employs systems thinking to appreciate and intervene in complex problematic situations (Bunch 2016; Parkes and Horwitz 2009). Ecohealth adopts system approaches to promote the health of people, animals and ecosystems in the context of social and ecological interactions (Parkes et al. 2014; Parkes and Horwitz 2016). This understanding has been strongly associated with 'ecosystem approaches to health' (also known as Ecohealth approaches) that link population (or, community) health and well-being with the environment and sustainable development (Lebel 2004; Charron 2012), and the field has expanded alongside the development of the journal *EcoHealth* which was founded in 2004 with aspirations toward a 'transdisciplinary imperative for a sustainable future' (Wilcox et al. 2004). The relevance and intersections of themes of health, environment and sustainability apply not only at the local level, but also at the regional, national and global levels, and, by extension, to the biosphere; and hence, the connection with the mission of the Global Ecological Integrity Group (GEIG).

Consistent with the body of work developed by IDRC around ecosystem approaches to health, this chapter uses the spelling of Ecohealth; where relevant, we may use *EcoHealth* (per the journal), or ecohealth to respect the specific uses or quotes by particular authors.

The development of Ecohealth

Ecohealth approaches are based on the understanding that health is an emergent property of inter-relationships among coupled human and natural systems (Bunch and Waltner-Toews 2015). Ecosystem approaches to health arose in the 1990s from a rich background of intellectual ferment as Canada wrestled with diverse problems ranging from Great Lakes contamination to zoonotic diseases. Canada's International Development Research Centre (IDRC) (Charron 2012; Forget and Lebel 2001; Lebel 2004) played a lead role in supporting an international community of scientists and scholars who advanced ecosystem approaches to health (Webb et al. 2010).

These collective efforts have enabled and, indeed, facilitated a shift to a research paradigm that embraces the six principles identified by Charron (2012):

- systems thinking;
- transdisciplinarity;
- participation;
- gender and social equity;
- sustainability; and
- knowledge-to-action.

Ecohealth is also a framework to study and manage relationships between human beings and the environment in pursuit of co-benefits that simultaneously improve ecosystem health and human well-being. It is an approach rooted in systems thinking and holism.

Examples of research using Ecohealth approaches

Ecohealth approaches have been applied in a range of different contexts, as exemplified by the series of international case studies examined as 'innovative applications' and profiled by Charron's 2012 text. A particularly striking interplay of the human manipulation of the natural environment came to public attention with the tsunami in Asia, especially Thailand, in 2004, when it was recognized that the removal of mangroves worsened the impact of the tsunami (Danielsen et al. 2005).

Other examples include a recent study by Omid et al. (2015), which showed that having 10 more trees in a city block improves health perception in ways comparable to:

- an increase in annual personal income of $10,000 and moving to a neighbourhood with $10,000 higher median income; or
- being 7 years younger.

A notable application of ecosystem approaches to health has contributions to understanding and responding to the linkages between watersheds (also known as catchments and river basins) and the determinants of health and well-being (Parkes et al. 2010; Morrison et al. 2017). Such approaches have been developed and applied across a range of countries and contexts including Australia, Canada, New Zealand and beyond (Parkes and Horwitz 2009; Parkes et al. 2010; Bunch et al. 2011). They have influenced consideration of the cumulative determinants of impacts in relation to land and water governance in the context of rapid social-ecological change, resource development and climate change (Parkes 2016; Picketts et al. 2016).

An active example of Ecohealth research in practice within this watershed context comes from the Credit Valley Conservation (CVC) Authority in Southern Ontario, Canada (Bunch 2016):

> CVC is a community-based environmental organization, dedicated to protecting, restoring and managing the natural resources of the Credit River Watershed in Ontario, Canada. Established by the provincial government in 1954, CVC is one of 36 Conservation Authorities in Ontario. As the primary scientific authority for the watershed, CVC works in partnership with municipal governments, schools, businesses and community organizations to deliver locally based programs.
>
> (Credit Valley Conservation Authority 2017)

This project was oriented to identifying and communicating relationships among ecosystem services and human health so as to demonstrate the importance of watershed management. The project used the Millennium Ecosystem Assessment framework that identified relationships among ecosystems (via ecosystem services) and various domains of human health and well-being to demonstrate such relationships in the local watershed context (Millennium Ecosystem Assessment 2005). For example, watershed residents interested in relationships among environment and obesity are presented both with information about access to green space, active lifestyles and obesity (including links to scientific studies) and also with maps of their local area that portray locations of nearby bike and walking trails, and local parks.

Continued work in scenario planning operationalizes this approach, using the ecosystem services cascade (Haines-Young and Potschin 2010) to assist the Conservation Authority in planning for conservation interventions based on understanding of the impacts on human health and well-being. These projects, typical of the Anthropocene, are necessary to reveal how human systems impact the natural systems upon which they depend.

Many Ecohealth projects have been undertaken in the Global South, in part thanks to funding and field-building contributions of Canada's IDRC (Cole et al. 2006; Charron 2012). One example of this, is work led by Dr Carlota Monroy and her team in Guatemala. Since 2004, they have worked with communities, anthropologists, sociologists, architects and engineers to apply an ecosystem approach to health for the control of Chagas disease. This has involved assessment and prioritization of risk factors for Chagas transmission, development of house improvement methods by using local materials and adapting traditional practices (Monroy et al. 2012). They have been successful in counteracting risk factors that favour the presence of Chagas disease vectors. In part, this involved working with educational and behavioural change in communities, and building capacity in the Ministries of Health in several countries to execute house improvements.

Several NGOs, such as World Vision, USAID, CARE, FAO-PESA and PRESANCAII, have adopted this approach to addressing social and ecological drivers of a notable infectious disease, helping to improve more than 4,000 houses in Central America, including the relocation of domestic animals (especially chickens) outside of homes. This has led to a reduction in the household of the vectors ('kissing bugs') that spread Chagas disease and a shift of blood sources from humans to chickens (Monroy et al. 2012). Geohelminth infections in children were also significantly reduced. Community engagement in these projects in Guatemala is essential, not only to aid in understanding the roles that local people play in their environments and the broader ecosystem, but also to increase their knowledge to better appreciate these connections. This reflects wider trends and commonalities in the application of Ecohealth as a field across multiple contexts (Saint-Charles et al. 2014).

All kinds of feedback loops are identified for these approaches, not only in identifying the interconnections throughout the environment, but also among the communities that are embedded in these ecosystems. For example, in the project with Credit Valley Conservation, community engagement was achieved through the conduct of health surveys and focus groups, as well as workshops with expert stakeholders. Similarly, in the Guatemala project, community participation and education were key factors for success. For example, in addition to house construction methods, Dr Monroy's team addressed domestic animal management with community members, such as with the construction of wire chicken coops, and the vaccination of chickens. This significantly decreased chicken death rates, increasing the number of chickens per household, consumption of meat, and it also generated additional income for women (Monroy et al. 2012).

Other examples demonstrate the relevance of Ecohealth approaches when considering impacts on indigenous communities whose traditional lands have been exploited for their resources. As the human enterprise and associated resource development expands, it encroaches on areas that were once the confine of communities that have lived off the land for centuries, with far-reaching implications for health and well-being of those affected (Gislason et al. 2017).

Ongoing encroachment of traditional lands for the exploitation of mineral resources, construction of hydroelectric dams and pipelines, offer a direct example of the need to understand the impacts on land and water resources in terms of their combined impacts on social, ecological, cultural determinants of health, and the potential to exacerbate existing health equities for populations already marginalized in rural and remote areas, while also generating long-term impacts on current and future generations (Mitchell-Foster and Gislason 2016; Parkes 2016).

In summary, Ecohealth approaches and the principles linked with them, encourage paying attention to diverse sources of exposure (both social and ecological) through the use of holistic frameworks and recognition of multiple, potentially interacting health outcomes. Of particular importance is the recognition that Ecohealth approaches foster understanding of influences on health and well-being from all perspectives, with direct and indirect pathways including cultural, social, economic and ecological impacts development.

The role of epidemiology in Ecohealth

Epidemiology is the science that bridges laboratory research with the human experience, the latter being deemed a 'natural experiment.' As an applied science, it provides the rational basis for informed policy decisions. Epidemiologists study health effects in relation to exposures to which people may be subjected in their lived experience.

The significance and relevance of ecosystems and Ecohealth approaches for epidemiology was highlighted by Soskolne et al. (2007). This work was further extended by Ladd and Soskolne (2008) where a toolkit comprising a suite of qualitative methods is provided and published in the 2008 GEIG book 'Sustaining Life on Earth: Environmental and Human Health through Global Governance.'

In these publications, the distinction is made between *Newtonian* and *complexity* paradigms as a mechanism for understanding the need to incorporate Ecohealth approaches and principles into epidemiological methods. The main thrust is to incorporate methods that move epidemiology from the purely *quantitative* approach (as in the Newtonian paradigm) to include *qualitative* methods needed to more fully address complexity through access to an Eco-Epidemiological toolkit. It is argued that this extension is needed if we are to embrace complexity and thereby identify the broader, more upstream determinants of health and well-being.

From the eco-epidemiological toolkit

As noted earlier, Ladd and Soskolne (2008) elaborated on a full 'toolkit for eco-epidemiological enquiry.' Here only the *integrated assessment* and *participatory methods* are addressed by way of example, highlighting areas of overlap and congruence with Ecohealth approaches and the principles and applications described above.

Integrated assessment

Rotmans (1998) and Rotmans and Dowlatabadi (1998) described integrated assessment as 'A structured process of dealing with complex issues, using knowledge from various scientific disciplines and/or stakeholders, such that integrated insights are made available to decision-makers.'

Integrated assessment is applied because:

- it provides a systems-based perspective;
- it provides a holistic and integrated (disciplinary) approach, providing perspectives for addressing global health issues;
- it places a greater emphasis on understanding processes (pathways) than on prediction (cause and effect);
- an interdisciplinary approach is required; and
- it makes explicit the multiple interactions that exist among natural, economic and social systems.

Participatory methods

With the seminal work of Functowicz and Ravetz (1994), participatory methods became recognized for their contribution, adding qualitative strength through a post-normal approach to science. Foundational to participatory methods are:

- policy exercises and focus groups; and
- the involvement of relevant stakeholders.

The participatory methods framework includes:

- focus groups that elicit preferences, opinions and viewpoints;
- participatory modelling, allowing stakeholders to explore the implications of their ideas;
- scientist-stakeholder workshops so that a research agenda can be formulated;
- identifying key issues, such that stakeholders can explore a range of possible futures; and
- conducting policy exercises wherein participants can assume different roles to simulate a decision-making process.

The development of One Health and the Manhattan Principles

In 2004, under the theme 'Building Interdisciplinary Bridges to Health in a "Globalized World"', a group of strategic thinkers met in New York City and formulated the 12 Manhattan Principles. The 12 Principles call on the international community to adopt a holistic approach to combat 'threats to the health of life on Earth' under the banner 'One World, One Health.' The latter has developed into a distinct and complementary approach known as One Health. It is its explicit and consistent orientation in linking human and animal health with increasing attention to ecological context (Zinsstag et al. 2011) that has led to frequent comparisons with Ecohealth approaches.

The meeting was organized by the Wildlife Conservation Society (WCS) and the Principles were disseminated through the US Centers for Disease Control and Prevention website. Held at Rockefeller University in New York City, the symposium set priorities for an international, interdisciplinary strategy for combating threats to the health of life on Earth.

The Manhattan Principles (2004) urge world leaders, civil society, the global health community and institutions of science to holistically tackle the prevention of epidemic/epi-zoonotic disease and the maintenance of ecosystem integrity. The first of the 12 Principles calls for recognizing the link between human, domestic animal and wildlife health, and the threat disease poses to people, their food supplies and economies, and the biodiversity essential to maintaining the healthy environments and functioning ecosystems we all require. The remaining Principles expand on ways to operationalize this Principle and implicitly includes Ecohealth approaches throughout.

Integration of all approaches: challenges and opportunities

There are clear parallels between Ecohealth and One Health approaches. Furthermore, both approaches align nicely with the Earth Charter (2000) as well as with the mission of the Global Ecological Integrity Group (GEIG; see www.globalecointegrity.org).

One Health has its focus on infectious/communicable diseases especially affecting human and animals, while Ecohealth has its focus on the totality of environmental determinants of health by combining both occupational, community-wide and naturally occurring exposures of concern. Both approaches rely heavily on quantitative as well as qualitative methods that are integrative and transdisciplinary; neither is reductionist, but take a systemic or holistic approach that embraces complexity. These synergies and overlaps mean that there is ongoing attention to the convergence between the two fields. Attention is focussed on recognizing distinct differences, some tensions, and ongoing areas of cross-fertilization between the two emerging fields. Differences include the fact that Ecohealth approaches tend not to give similar levels of attention to the health of non-human species, and that One Health approaches have tended to less fully embrace the social *and* ecological context of human and animal health (Zinsstag et al. 2011; Zinsstag 2012; Mi et al. 2016).

A methodological caution

In epidemiological research, the control group provides the reference point against which the experimental (or, exposed) group's health experience is compared. The Ecohealth approach could be affected by bias. This can arise by virtue of the fact that it is not always possible to identify a purely unexposed control group, especially in the context where a small indigenous community is being studied. Where control groups are selected from among the broader community that may itself be exposed – albeit to low-levels of contaminants – the relative differences found in case-control or cohort studies will be lower than they ought to be. In these circumstances, we will observe an attenuation towards the null hypothesis as a bias built into the design of the study (i.e., demonstrating less of an effect than what, in truth, exists).

Conclusions

The need to address complexity necessitated the development of post-normal approaches to science. These approaches were elevated through, among others, the foresight of Canada's IDRC, who has had a significant role in the emergence of ecosystem approaches to health. The emergence of One Health represents a communicable and emerging infectious disease thrust, heavily influenced by zoonotic pathogens. With complexity being embraced in many areas of enquiry, there is considerable overlap in both rationale and approach. Epidemiology's role is to integrate the qualitative methods and bring its own rigour to evaluate bias to the fore.

Acknowledgements

It is on the shoulders of others that we stand, like those of Dominique Charron, Jean Lebel, Pim Martens, Anthony (Tony) J. McMichael, Donna Mergler, David Rapport, Jan Rotmans, Jerry Spiegel, David Waltner-Toews, Annalee Yassi and others within Canada's International Development Research Centre (IDRC).

References

Bavington, D. (2002) 'Managerial Ecology and Its Discontents: Exploring the Complexity of Control, Careful Use and Coping in Resource and Environmental Management'. *Environments*, vol 30, no 3, pp3–21.

Boyd, D. R. (2012a) *The Environmental Rights Revolution: A Global Study of Constitutions, Human Rights, and the Environment*. Vancouver: UBC Press.

Boyd, D. R. (2012b) *The Right to a Healthy Environment: Revitalizing Canada's Constitution*. Vancouver: UBC Press.

Bunch, M. J. (2016) 'Ecosystem Approaches to Health and Well-Being: Navigating Complexity, Promoting Health in Social–Ecological Systems'. *Systems Research and Behavioral Science*, vol 33, no 5, pp614–632. https://doi.org/10.1002/sres.2429

Bunch, M. J., Morrison, K. E., Parkes, M. W. and Venema, H. D. (2011) 'Promoting Health and Well-being by Managing for Social-Ecological Resilience: The Potential of Integrating Ecohealth and Water Resources Management Approaches'. *Ecology and Society*, vol 16, no 1, 6, available at www.ecologyandsociety.org/vol16/iss1/art6.

Bunch, M. and Waltner-Toews, D. (2015) 'Grappling with Complexity: The Context for One Health and the Ecohealth Approach'. In J. Zinsstag, E. Schelling, D. Waltner-Toews, M. Wittakerand and M. Tanner (eds), *One Health: The Theory and Practice of Integrated Health Approaches* (pp415–426). Boston, MA: CABI.

Butler, C. D. (2013) 'Tibet, China, Protest Immolation and Social Medicine'. In M. K. Gislason (ed.), *Ecological Health: Society, Ecology and Health* (pp67–89). Bingley: Emerald Press.

Charron, D. E. (ed.). (2012) *Ecohealth Research in Practice: Innovative Applications of an Ecosystem Approach to Health*. Ottawa, Ontario: Springer and International Development Research Centre.

Cole, D. C., Crissman, C. and Orozco, F. (2006) 'Canada's International Development Research Centre's Eco-Health Projects with Latin Americans'. *Canadian Journal of Public Health*, 97(6), I8–I14.

Credit Valley Conservation Authority. (2017) 'About CVC'. Available at www.creditvalleyca.ca/about-cvc (retrieved 11 October 2017).

Danielsen, F., Sørensen, M. K., Olwig, M. F., Selvam, V., Parish, F., Burgess, N. D., Hiraishi, T., Karunagaran, V. M., Rasmussen, M. S., Hansen, L. B., Quarto, A., and Suryadiputra, N. (2005) 'The Asian Tsunami: A Protective Role for Coastal Vegetation'. *Science*, vol 310, p643.

Earth Charter (2000) 'The Earth Charter'. Available at http://earthcharter.org/invent/images/uploads/echarter_english.pdf.

Forget, G. and Lebel, J. (2001) 'An Ecosystem Approach to Human Health'. *International Journal of Occupational and Environmental Health*, 7(2), S1–S36.

Funtowicz, S. O. and Ravetz, J. R. (1994) 'Uncertainty, Complexity and Post-normal science'. *Environ Toxicol Chem*, vol 13, pp1881–1885.

Gislason, M. K., Buse, C., Tosh, J., Woollard, R. W., Parkes, M. W. and Griffin Cohen, M. (2017) 'The Complex Impacts of Intensive Resource Extraction on Women, Children and Aboriginal Peoples: Towards Contextually-Informed Approaches to Climate Change and Health'. Chapter 14 in. In *Climate Change and Gender in Rich Countries: Work, public policy and action*. Abingdon: Earthscan from Routledge.

Haines-Young, R. and Potschin, M. (2010) 'The Links between Biodiversity, Ecosystem Services and Human Well-being'. In D. Raffaelli and C. Frid (eds), *Ecosystem Ecology: A New Synthesis* (pp110–139). Cambridge: Cambridge University Press.

Hallström, L. K., Guehlstorf, N. and Parkes, M. W. (2015) 'Convergence and Diversity: Integrating Encounters with Health, Ecological and Social Concerns' (pp3–28). Chapter 1. In L. K. Hallström, N. P. Guehlstorf and M. W. Parkes (eds), *Ecosystems, Society and Health: Pathways through Diversity, Convergence and Integration*. Montreal, Quebec,McGill Queens University Press.

Hofferth, S. L. and Sandberg, J. F. (2001) 'How American Children Spend Their Time'. *Journal of Marriage and Family*, vol 63, no 2, pp295–308. https://doi.org/10.1111/j.1741-3737.2001.00295.x

Ladd, B. D. and Soskolne, C. L. (2008) 'A Toolkit for Eco-epidemiological Enquiry under Global Ecological Change'. In C. L. Soskolne, L. Westra, L. J. Kotzé, B. Mackey, W. E. Rees and R. Westra (eds), *Sustaining Life on Earth: Environmental and Human Health through Global Governance* (pp369–382). Lanham, MD: Lexington Books.

Lebel, J. (2004) 'Ecohealth and the Developing World'. *EcoHealth*, vol 1, no 4, pp25–26.

Louv, R. (2005) *Last Child in the Woods: Saving Our Children from Nature-Deficit Disorder*. Chapel Hill, NC: Algonquin Books.

Malone, K. (2007) 'The Bubble-Wrap Generation: Children Growing Up in Walled Gardens'. *Environmental Education Research*, vol 13, no 4, pp513–527. https://doi.org/10.1080/13504620701581612

Manhattan Principles (2004) 'The Manhattan Principles: As Defined during the Meeting Titled Building Interdisciplinary Bridges to Health in a "Globalized World" Held in 2004'. Available at www.cdc.gov/onehealth/pdfs/manhattan/twelve_manhattan_principles.pdf.

McMichael, A. J. (1999) 'Prisoners of the Proximate: Loosening the Constraints on Epidemiology in an Age of Change'. *American Journal of Epidemiology*, vol 149, no 10, pp887–897.

Mi, E., Mi, E. and Jeggo, M. (2016) 'Where to Now for One Health and Ecohealth?' *EcoHealth*, vol 13, no 1, pp12–17. https://doi.org/10.1007/s10393-016-1112-1

Millennium Ecosystem Assessment. (2005) *Ecosystems and Human Well-being: Synthesis, A Report of the Millennium Ecosystem Assessment*. Washington, DC: Island Press. Available at www.millenniumassessment.org/documents/document.356.aspx.pdf.

Mitchell-Foster, K. and Gislason, M. K. (2016) 'Lived Reality and Local Relevance: Complexity and Immediacy of Experienced Cumulative Long-Term Impacts'. Vignette 6, in Chapter 6, 'Exploring Cumulative Effects and Impacts Through Examples'. In M. P. Gillingham, G. R. Halseth, C. J. Johnson and M. W. Parkes (eds), *The Integration Imperative: Cumulative Environmental, Community and Health Impacts of Multiple Natural Resource Developments* (pp173–175). Cham: Springer International Publishing AG.

Monroy, C., Castro, X., Bustamante, D. M., Pineda, S. S., Rodas, A., Moguel, B. and Quiñonez, J. (2012) 'An Ecosystem Approach for the Prevention of Chagas

Disease in Rural Guatemala'. In *Ecohealth Research in Practice* (pp153–162). New York: Springer.

Morrison, K., Bunch, M. J. and Hallström, L. (2017) 'Public Health at the Watershed Scale'. In S. Renzetti and D. P. Dupont (eds), *Water Policy and Governance in Canada* (pp337–356). Cham: Springer International Publishing. https://doi.org/10.1007/978-3-319-42806-2_18

Omid K., P. Gozdyra, B. Misic, F. Moola, L. Palmer, T. Paus and M. Berman (2015) 'Neighborhood Greenspace and Health in a Large Urban Center'. *Scientific Reports*, vol 5. doi:10.1038/srep11610

Parkes, M. W. (2016) 'Cumulative Determinants of Health Impacts in Rural, Remote, and Resource-Dependent Communities'. In M. P. Gillingham, G. R. Halseth, C. J. Johnson and M. W. Parkes (eds), *The Integration Imperative: Cumulative Environmental, Community and Health Impacts of Multiple Natural Resource Developments* (pp117–152). New York: Springer International Publishing.

Parkes, M. W. and Horwitz, P. (2009) 'Water, Ecology and Health: Ecosystems as Settings for Promoting Health and Sustainability'. *Health Promotion International*, vol 24, no 1, pp94–102. https://doi.org/10.1093/heapro/dan044

Parkes, M. W. and Horwitz, P. (2016) 'Ecology and Ecosystems as Foundational for Health'. In H. Frumkin (ed.), *Environmental Health: From Global to Local* (3rd edn). San Francisco, CA: Jossey-Bass.

Parkes, M. W., Horwitz, P. and Waltner-Toews, D. (2014) 'Ecohealth'. In A. C. Michalos (ed.), *Encyclopedia of Quality of Life and Well-Being Research*. Heidelberg: Springer-Verlag.

Parkes, M. W., Morrison, K. E., Bunch, M. J., Hallström, L. K., Neudoerffer, R. C., Venema, H. D. and Waltner-Toews, D. (2010) 'Towards Integrated Governance for Water, Health and Social-Ecological Systems: The Watershed Governance Prism'. *Global Environmental Change*, vol 20, pp693–704.

Picketts, I. M., Parkes, M. W. and Déry, S. J. (2016) 'Climate Change and Resource Development Impacts in Watersheds: Insights from the Nechako River Basin, Canada'. *The Canadian Geographer / Le Géographe Canadien*, vol 61, no 2, pp196–211. https://doi.org/10.1111/cag.12327

Rotmans, J. (1998) 'Methods for IA: The Challenges and Opportunities Ahead'. *Environmental Modeling and Assessment*, vol 3, pp155–179.

Rotmans, J. and Dowlatabadi, H. (1998) 'Integrated Assessment Modeling'. In: *The Tools for Policy Analysis*. (Raynes S, Malone E, eds). Columbus, OH: Battelle Press, pp291–377.

Saint-Charles, J., Webb, J., Sanchez, A., Mallee, H., van Wendel de Joode, B. and Nguyen-Viet, H. (2014) 'Ecohealth as a Field: Looking Forward'. *EcoHealth*, vol 11, no 3, pp300–307. https://doi.org/10.1007/s10393-014-0930-2

Soskolne, C. L., Huynen, M., Ladd, B. D. and Martens, P. (2007) 'Eco-epidemiological Enquiry under Global Ecological Change: An Integrated Assessment Toolkit For Beginners'. 25th Anniversary of the Collegium Ramazzini, Carpi, Italy. Available at www.colinsoskolne.com/documents/change.pdf.

Webb, J., Mergler, D., Parkes, M. W., Saint-Charles, J., Spiegel, J., Waltner-Toews, D., Yassi, A. and Woollard, R. F. (2010) 'Tools for Thoughtful Action: The Role of Ecosystem Approaches to Health in Enhancing Public Health'. *Canadian Journal of Public Health*, vol 101, pp439–441.

Wilcox, B., Aguirre, A. A., Daszak, P., Horwitz, P., Martens, P., Parkes, M., Patz, J. and Waltner-Toews, D. (2004) 'EcoHealth: A Transdisciplinary Imperative for a Sustainable Future'. *EcoHealth*, vol 1, pp3–5.

Zinsstag, J. (2012) 'Convergence of Ecohealth and One Health'. *EcoHealth*, vol 9, no 4, pp371–373. https://doi.org/10.1007/s10393-013-0812-z

Zinsstag, J., Schelling, E., Waltner-Toews, D. and Tannera, M. (2011) 'From "One Medicine" to "One Health" and Systemic Approaches to Health and Well-being'. *Preventive Veterinary Medicine*, vol 101, pp148–156.

9 Trading health

A community health impact assessment perspective of the Trans-Pacific Partnership

Robert Rattle and Laura Tomie

Introduction and background

There are few tools available to understand the effects of international trade agreements on the health and well-being of populations. This has left a key policy and negotiating gap. One tool used to evaluate trade agreements is a health impact assessment (HIA). HIA is a combination of procedures, methods and tools that systematically judges the potential effects of a policy, programme or project on the health of a population and the distribution of those effects within the population (Rattle and Kwiatkowski 2003). The HIA process involves five major steps: screening, scoping, assessment and evaluation, recommendations and reporting, and monitoring and follow-up.

HIAs of trade agreements generally take a state level or multilateral approach to potential health impacts. Accordingly, results typically identify national and international responses to improve the health outcomes of these agreements. While these approaches conform with the policy level within which trade agreements are negotiated, they offer limited information to local constituencies for understanding how proposed international trade agreements will affect local populations. Further, the detail that illuminates how different communities and constituencies will be affected by trade agreements is forfeit in national averages. The result is an information and resources gap for local decision makers with which to balance the adverse impacts and leverage the beneficial potentials of international trade agreements. This is further aggravated by broad policy responses which often overlook social and environmental injustices inflicted upon at risk populations under the mantra of economic growth.

To fill these critical gaps, Crane undertook an HIA of the Trans-Pacific Partnership (TPP) for Sault Ste. Marie, Ontario. Sault Ste. Marie is located in the Algoma district of Northern Ontario. With a population of 73,368, it is the third largest city in Northern Ontario. The city is situated on the border between the United States and Canada, and with the St. Mary's River, the Sault Locks, and the Trans-Canada highway, it is a busy location for shipping and transportation. The population demographics exhibit some unique characteristics different from the average for the rest of Canada (identified below), and even the province of Ontario, however, not necessarily unique from many other national or provincial outliers. Sault Ste. Marie is serviced by Algoma Public Health which services the entire Algoma District of Northern Ontario.

The TPP is an international trade agreement of 12 original signatories that border the Pacific Ocean. The countries are Australia, Brunei, Canada, Chile, Japan, Malaysia, Mexico, New Zealand, Peru, Singapore, United States and Vietnam. At the time of this study, the former Obama administration of the United States had entered into the partnership, however as of January 2017 the United States have withdrawn. The partnership aims to remove and/or lower tariff barriers to trade, and essentially create new policies in regards to international trade between member countries.

This study applied a desk-based approach to assess health impacts from the TPP on the population health of the city of Sault Ste. Marie. There is a lack of local and regional studies that look at how international trade deals affect smaller cities. By bringing impacts down to a local level, municipal and regional stakeholders, citizens, and all levels of government can better understand how municipal characteristics may affect population health outcomes from international trade agreements. All data used was secondary, and as data was occasionally only available at a district level, numbers taken from the Algoma District were used as proxy indicators for Sault Ste. Marie.

The HIA applied local data in key relevant areas identified by other HIAs of trade agreements and those of the TPP agreement (Labonté et al. 2016; Hirono et al. 2015) to better understand the positive and adverse effects of the proposed TPP agreement on the local population. Local data evaluated included health and other social and environmental data. Community characteristics such as income, demographics, geography, and education levels were found to be important determinants affecting local health outcomes from the TPP.

When discussing potential health impacts of major trade deals, consideration must be given to the priority populations or vulnerable groups that will be more sensitive to the negative effects established. While the average age of 44.6 is not vastly different than the Canadian average of 41, the percentage of the population 65 years of age and older in 2014 in Canada was 16.9 per cent, while in Sault Ste. Marie it was 22.4 per cent and projected to rise to 30 per cent by 2018 (Sault Ste. Marie 2014; Statistics Canada 2011).

The average after tax income in Sault Ste. Marie is $32,374 compared with $34,284 for Canada (Statistics Canada 2011). Having a lower income is a well-documented social determinant of health that already leaves individuals with a lower chance of achieving optimal health.

Sault Ste. Marie is also characterized by an urban Indigenous population of 9.8 per cent compared with a national average of 3.8 per cent and provincial average of 2.0 per cent (Statistics Canada 2006). Indigenous children and youth are more likely to live in a single parent household compared with their non-Indigenous peers, and less likely to be attending school (66 per cent versus 72 per cent). The Indigenous unemployment rate was 11.9 per cent compared with 7.4 per cent for the non-indigenous population, and Indigenous groups are experiencing a widening earnings gap with lower incomes for both women and men compared with their non-Indigenous counterparts ($17,000/$18,000 versus $20,000/$34,000). Compared with Canada, income levels, dual parent households, and school attendance remain lower for Sault Ste. Marie.

This HIA illustrates a sample of the many health impacts that a community with similar demographics to Sault Ste. Marie might confront. The health impacts from tobacco and alcohol consumption, access to pharmaceuticals, nutritional and dietary choices, and environmental effects resulting from the proposed TPP are described below.

Tobacco

Certain chapters in the TPP may lead to a decreased governmental role in the regulation of tobacco products. There have been many advances to reduce the consumption and marketing of tobacco products. The rules about indirect expropriation, which states that items such as trademarks and logos are intellectual property of the manufacturing company, was found to increase the potential to challenge national health legislation (Hirono et al. 2015). This suggests that the TPP may encourage tobacco companies to dispute plain packaging legislation. Furthermore, the Regulatory Coherence Chapter articulates that governments are required to provide opportunities for stakeholder input into policy-making. This could potentially undermine the WHO's Framework Convention on Tobacco Control, which states that tobacco control policies be protected from tobacco industry interests. Both of these chapters plus the overall increase of cross-border services suggests reduced health outcomes from the TPP and challenges under certain community characteristics.

In the Algoma District, the number of smokers (25.7 per cent) is higher than that of Ontario (18.6 per cent) (Statistics Canada 2009; APH 2011). The lung and bronchus cancer rates in Sault Ste. Mari are 64.7 per 100,000 versus 52.5 per 100,000 for Ontario, with mortality rates also higher than Ontario at 52.1 per 100,000 versus 41.4 per 1000 (APH 2015a). With higher than average tobacco consumption rates in the Algoma District, the TPP may contribute to, or re-introduce, a tobacco consumption normalization that would militate against denormalization efforts of federal, provincial and district policies and programmes. This may stress already limited public health resources and the efficacy of tobacco cessation programmes and policies. The most vulnerable groups are those of low socio economic status, indigenous peoples, current drug users, and adolescents.

Alcohol

Alcohol is currently not subjected to the same regulatory standards as are tobacco products. The laws for marketing and sales are more relaxed, and many warning labels, such as those for drinking while pregnant, are optional. As with tobacco products, certain TPP chapters may impede on the government's ability to regulate and execute regulatory procedures relevant to the marketing and availability of alcoholic beverages. Three chapters in particular prove to be troublesome. The Intellectual Property chapter, as with tobacco products, would give alcohol companies trademark rights to logos and other marketing tools without governmental regulation. The Cross-Border Services chapter indicates that governments may have less control over introducing limits on the number and size of services supplied across borders. This could affect attempts to restrict the number of licenced alcohol outlets per geographic area. The Investor State Dispute Settlement (ISDS) may allow alcohol companies to take legal action against governments over policy decisions that adversely impact investments.

The overarching theme of these chapters is that regulation and control of the marketing and sales of alcohol products could be severely limited in terms of governmental role, and companies themselves would be given much more power.

Short term risks of alcohol consumption include injuries, alcohol poisoning, and negative effects on the foetus of pregnant women. Long-term risk of excessive alcohol use may lead to chronic diseases and serious illnesses such as high blood pressure, many types of cancer, psychological disorders, social problems, and alcoholism (CDC 2016). Those of a higher socio-economic status tend to engage in high-risk drinking more than those of a low socio-economic status. However, the lower the socio-economic status, the higher the rates of mortality and disability as a result of high risk drinking. One issue that could be seen as a result of changes in policies surrounding alcohol, particularly in the Cross-Border Services chapter, implies less government intervention regarding the density of alcohol retailers in a certain geographic location. A study by Popova et al. (2009) outlines an increase in all negative health effects of alcohol consumption when outlets are more densely placed. Having more licenced alcohol retailers will increase the amount of marketing and the availability of alcoholic products to the population. Once again, the normalization of this practice would function against policies and programmes to reduce excess alcohol consumption.

One group in particular that is susceptible to marketing is adolescents. In Sault Ste. Marie, almost 50 per cent of students in grades 7–12 reported drinking alcohol in the past 12 months, with around 20 per cent reporting binge drinking during the same time period (APH 2015b). Additionally, oesophageal cancer which is linked to alcohol usage is higher in Algoma (6.2 per cent) than Ontario (4.1 per cent) (APH 2015a). Reduced resources or an increased resource need in health policy and practice will challenge existing resource allocations, especially in areas already underserviced and those with high population risk factors and higher than average health risk vulnerabilities. Overall, a decrease in government involvement in the regulation of marketing and distribution of alcohol suggests an antagonistic environment for health policy and practice in general and an adverse impact in Sault Ste. Marie. The result may be an adverse health impact on population health in Sault Ste. Marie from alcohol consumption practices derived from the TPP.

Food labelling

The TPP has also been criticized for its treatment of food labelling, especially through the ISDS mechanism. Within the ISDS, the food manufacturing industry has the ability to take legal action against state regulatory efforts. Additionally, the regulatory coherence and transparency chapters outline that the processed food industry would have a greater role in policy decisions, and thus would be able to influence the type of food labelling system used. Furthermore, the chapter on Technical Barriers to Trade (TBT) outlines that policies and products need to be 'not more trade-restrictive than necessary to fulfil a legitimate objective'. As this statement is open to interpretation, measures could be put in place which prevent innovations to assist consumers to make better food choices. The processed food industry could be compelled to highlight the very small nutritional value of their products, and try to conceal the negative elements of the product in order to increase consumption. Finally, as the common theme of the TPP is to ease trade barriers between member countries, cross border advertising clauses may make it much more difficult to regulate marketing of foods to children. In Canada, The Ottawa Principles created in 2016 were an effort from the Heart Stroke Foundation, Childhood Obesity Foundation and the Stop Marketing to Kids Coalition in order to restrict food and beverage marketing to kids. Advocacy efforts such as these will more than likely be hindered by inconsistent international marketing laws.

The health impacts of inconsistent and non comprehensive food labels make choosing healthier food more challenging for the consumer. In particular, older adults, and culturally and linguistically diverse groups will find the lack of nutritional support decision making tools increasingly confusing under global trade in food products. This is due to potential language barriers, and non accessible labels in terms of ease of comprehension and font choice.

Janssen et al. (2006) suggest that there is a strong correlation between socioeconomic status, obesity, unhealthy eating, and physical inactivity in Canadians. The overweight/obese rates of adults 18 and over in Algoma are approximately 63.7 per cent, versus 51.4 per cent in Ontario (APH 2011).

A total of 38 per cent of the population aged 25–64 do not possess post-high school education (Statistics Canada 2011). Given the lack of standards defined for a minimum literacy competency level for food labels, lower literacy rates may contribute to a greater number of unhealthy choices. Language is also a potential barrier to choosing healthy food options. While a majority of residents of Sault Ste. Marie speak at least one of the two official national languages, 8.6 per cent of the population are immigrants (Statistics Canada 2011). The generalized measures put in place by the TPP fail to provide adequate guidelines about food labels information or formats. Combined with marketing components of the TPP, the food industry is poised to pursue barrier-free food choices that reflect a high demand for their products with potential adverse impacts on population health in Sault Ste. Marie.

Pharmaceuticals

In many member nations there is already a very high, and steadily increasing, out of pocket cost for medicines. As with the other health impacts discussed above, pharmaceutical products will also be subjected to the same intellectual property, patent and investment rules. With decreased governmental involvement in the production and distribution of medicines, large pharmaceutical companies will obtain the power to acquire multiple patents for drugs and create a monopoly of ownership. As smaller, regional companies would not be able to compete, less competition will most likely lead to an increase in cost with the potential for decreased access, especially for those without private health insurance, and a reduced compensation programme for provincial health ministries.

An increased cost for pharmaceuticals has been shown to decrease adherence to medicine compliance, which results in a decline in overall population health. With a decline in population health comes increased hospitalizations and higher mortality rates. While this will affect everyone in a population, not all populations will be affected equally. Those with a low socio–economic status, the elderly and those on fixed incomes, and precariously employed individuals – all higher averages in Sault Ste. Marie – will experience an increased health burden. This will cause an increased demand on the public healthcare system, and could potentially effect wait times for procedures, accessibility to a family physician, drug and procedure coverage, and stress on available resources. Vulnerable groups such as those with chronic conditions, lower and fixed incomes are especially at risk of decreased access to pharmaceuticals.

Environmental impacts

Any environmental impacts will also have direct and indirect effects on human health. The TPP does not have comprehensive environmental guidelines or accountability mechanisms to define state level responsibilities for mitigating the increase of pollution from the production, use, disposal and transportation of goods. Rather than encouraging adoption of high environmental responsibility across all member countries, each state will determine their own level of domestic environmental protection and priorities. As no particular environmental standards are laid out, they are virtually unenforceable. Since the TPP aims to reduce trade barriers between nations, implementing high levels of environmental restrictions would potentially interfere with trade and investment. Clearly, the TPP has taken the classic approach to environmental externalities by treating pollution as external to economies. Accordingly, some level of environmental impact will occur, and each nation being responsible for their own policies creates a vacuum of high environmental standards and integration to complex ecological challenges.

With new shipping infrastructure being considered for Sault Ste. Marie and current shipping and road transport poised to increase, additional global transport of goods into, out of and through the community is likely to increase transportation related environmental pollution. This is likely once again to adversely affect the most vulnerable populations such as those currently living in low socio-economic neighbourhoods and Indigenous populations near proposed port and road infrastructure.

The risk of spills, ballast water dumping (and exporting) invasive species, and direct emissions lead to numerous environmental health concerns. Furthermore, an increase in shipping, one of the world's biggest contributors to GHG emissions, will contribute to accelerated climate change. Already exposed to numerous risks, including relative isolation, limited transportation options, low food security, water borne and vector borne diseases, and flooding, Northern Ontario and Sault Ste. Marie will be affected greater by climate change, with fewer resources for adaptation and mitigation. Further, Sault Ste. Marie remains financially stressed to meet provincial recycling and waste diversion standards due largely to the distance to markets and costs of transportation. An increase in global trade in consumer products with lower materials standards may lead to an increased stress on available community resources to recycle and divert waste, along with an increase in toxic waste from product disposals. At the other end, an increase in demand for consumer products through reduced trade barriers and consumer costs may lead to greater resources exploitation, potentially benefitting resource sectors (with the caveat of global pressures on industry to reduce costs) while stressing the ecological health of the District (Chong 2014; Ecojustice 2014).

With the increase of just-in-time delivery and e-commerce/internet shopping, combined with low population densities throughout Algoma and Northern Ontario, many goods are shipped by commercial truck on roads. This has increased truck traffic along the only major Trans-Canada Highway corridor through the region, raising public safety risks from traffic collisions, amplifying pollution, contributing to spill risks, air pollution and GHG emissions, severing migration routes, and causing economic impacts from highway closures due to collisions and weather-related events. With limited rail connectivity and a recently drafted Northern Ontario Multi-Modal Transportation Study (MTO/MNDM 2017) that offers no support for rail infrastructure (CAPT 2017), these trends are expected to continue, increasing the impacts from rising commercial truck traffic.

Discussion and conclusions

This HIA demonstrates, in particular, three key findings:

1. The TPP will adversely affect population health in Sault Ste. Marie due to specific community health states, characteristics and vulnerabilities.
2. Population health outcomes from international trade agreements vary between communities, depend on specific community characteristics, and can be important health determining factors.
3. A locally focused HIA can inform trade negotiations, and alert decision makers to key policy and programme needs.

Despite the relevance of trade for health, state level trade negotiators often overlook the impact on population health and are unable to evaluate community level impacts during negotiations. While this may have deleterious effects for their representative state, they may also have adverse impacts on foreign states by contradicting official policy (such as foreign aid). While several HIAs of international trade agreements exist, they do not generally include local data and the effects on local communities.

While it is typically the state which negotiates international trade agreements, community data is important to inform these negotiations and to help guide related state policies and programme developments.

This study suggests that, without local details, the distribution of international trade impacts on population health will be conflated with averages and medians, leaving some communities better off while others will suffer greater health impacts and inequities. This study also emphasized the importance of local data to support and understand the potential health impacts from international trade agreements. While there has been considerable movement towards open data sets, progress is very heterogeneous between and within states.

References

APH (2015). *Algoma Cancer Report 2015*. Sault Ste. Marie: Algoma Public Health.

APH (2011). *Chronic Disease Prevention Health Status Report 2010*, February 2011. Sault Ste. Marie: Algoma Public Health.

APH (2015b). *Alcohol and Youth*. December 2015. Sault Ste. Marie: Algoma Public Health.

CAPT (2017). Coalition for Algoma Passenger Trains' Response to the DRAFT 2041 Northern Ontario Multimodal Transportation Study. Available at http://captrains.ca/wp-content/uploads/2010/01/CAPTs-response-to-Draft-2041-NOMTS-sept-1417.pdf.

CDC (2016). Alcohol Use and Your Health – Fact Sheet. Centers for Disease Control and Prevention. Available at www.cdc.gov/alcohol/fact-sheets/alcohol-use.htm.

Chong, Jed (2014). *Resource Development in Canada: A Case Study on the Ring of Fire. Publication No. 2014-17-E31*. Ottawa: Economics, Resources and International Affairs Division, Parliamentary Information and Research Service.

Ecojustice (2014). *Getting it Right in Ontario's Far North: The Need for a Regional Strategic Environmental Assessment in the Ring of Fire. A Working Paper prepared by Cheryl Chetkiewicz and Anastasia M. Lintner*. Vancouver: Ecojustice.

Hirono, K., Haigh, F., Gleeson, D., Harris, P. and Thow, A. M. (2015). *Negotiating Healthy Trade in Australia: Health Impact Assessment of the Proposed Trans-Pacific Partnership Agreement*. Liverpool, NSW: Centre for Health Equity Training Research and Evaluation, Centre for Primary Health Care and Equity, Faculty of Medicine, UNSW Australia.

Janssen, Ian, Boyce, William F., Simpson, Kelly, and Pickett, William, (2006). Influence of Individual- and Area-Level Measures of Socioeconomic Status on Obesity, Unhealthy Eating, and Physical Inactivity in Canadian Adolescents. *American Journal of Clinical Nutrition* 83(1): 139–145.

Labonté, Ronald, Schram, Ashley, and Ruckert, Arne (2016). The Trans-Pacific Partnership: Is It Everything We Feared for Health? *International Journal of Health Policy and Management* 5(10): 1–10.

MTO/MNDM (2017). Northern Ontario Multimodal Transportation Strategy, Draft 2041. Northern Ontario Multimodal Transportation Strategy, September 2017. Available at: https://nomts.ca.

Popova, S., Giesbrecht, N., Bekmuradov, D. and Patra, J. (2009). Hours and Days of Sale and Density of Alcohol Outlets: Impacts on Alcohol Consumption and Damage: A Systematic Review. *Alcohol and Alcoholism* 44(5) (September–October).

Rattle, Robert and Kwiatkowski, Roy (2003). Defining Boundaries: Health Impact Assessment and Social Impact Assessment. In *International Handbook of Social Impact Assessment: Conceptual and Methodological Advances*, edited by Henk Becker and Frank Vanclay. Cheltenham: Edward Elgar.

Sault Ste. Marie (2014). *Age Friendly Sault Ste. Marie: Community Assessment Report.* September 2014. Sault Ste. Marie.

Statistics Canada (2006). *National Household Survey*. Ottawa: Statistics Canada.

Statistics Canada (2009). *Canadian Community Health Survey*. Ottawa: Statistics Canada.

Statistics Canada (2011). *National Household Survey*. Ottawa: Statistics Canada.

10 Indirect health effects consequent to the Fukushima nuclear accident of 11 March 2011

Yuliya Lyamzina

Introduction

According to the International Nuclear Event Scale (INES), the second biggest nuclear power plant accident in history occurred at the Fukushima Daiichi Nuclear Power Station (FDNPS) (IAEA 2015), on 11 March 2011. The combination of a tsunami, an earthquake and a nuclear accident has exposed confusion and shortcomings in emergency preparedness and response. The occurrence of such a compound disaster was far beyond what had ever been prepared for as part of the first response to a nuclear emergency (Tanigawa and Chhem 2014).

Procedures for medical response to a nuclear emergency have been learned from past experiences, such as the 1986 Chernobyl accident and, as a result, several compulsory radiation protection measures took place after the March 2011 events for both public and environmental protection, like sheltering, partial stable iodine administration (Hasegawa 2013), food and water restrictions. Evacuation orders were issued by the Government of Japan with gradual expansion covering in total 11 municipalities with a total population of about 165,000 people (Kingston 2012; Fukushima Prefectural Government 2017). According to the Fukushima Prefectural Government, a total of 164,865 people (peaked in May 2012) left their homes because of the nuclear accident following the Government's evacuation order, or by their own choice as a precautionary measure (Fukushima Prefectural Government 2017).

Despite the fact that the Fukushima accident is classified as Level 7 ('Major Accident') according to the INES of the International Atomic Energy Agency (IAEA) and the severity of the reactor damage, a number of post-accident assessment reports from national, international and non-governmental organizations confirm that, overall, environmental contamination from the Fukushima is less severe than the Chernobyl (IAEA 2015; Steinhauser, Brandl and Johnson 2014; Steinhauser and Saey 2016). Resulting from extensive environmental decontamination activities and natural radioactive decay, significant reductions in radiation levels are currently observed and confirmed by different sources. Therefore, the Government of Japan has been gradually lifting evacuation orders from those contaminated areas where the decontamination and remediation activities of the main infrastructure are fully completed (Ministry of the Environment 2017; Morino, Ohara and Nichizawa 2011).

As with any disaster, the three biggest nuclear accidents (i.e. Three Mile Island, Chernobyl, and Fukushima), have one health response in common for those who survive, namely severe psychological effects. These three events emphasize the importance of recognizing the human response to a disaster. Nuclear disasters, however, are a special case: they differ from other disasters in that they are not one-hit events; the effects of lingering exposure to radiation along with the longer-term concern about the consequences of having been exposed, makes a nuclear disaster quite different from a disaster that has only direct impacts that are immediate. Hence, adverse impacts of nuclear accidents are not limited to direct radiation-related health impacts; indirect health effects (other health and psycho-social issues) must be acknowledged and incorporated into international disaster management guidelines. Indirect health effects of radiation have complex psycho-social consequences, such as fear, chronic stress, anxiety and risk perceptions related to radiation leading to mental health problems, resulting from the loss of jobs, damaged family relationships and community bonds. These arise from constant changes in the social environment and an overall decrease in the quality of life (Becker 2013; Maeda and Oe 2017).

Currently, 24 studies are available demonstrating emotional and physical challenges during and immediately after the Fukushima disaster. These have resulted in a post-traumatic response. The effects of radiation-related anxiety and risk perception manifest as severe distress, resulting in the intention to leave employment and not to return home (Takebayashi, Lyamzina, Suzuki and Murakami 2017). Additionally, specific radiation issues such as 'radiation stigma' targeting evacuees have emerged and became a serious social issue after the disaster. Maeda and Oe (2017) even discuss that weakening community bonds due to the disaster may reduce resilience for overcoming challenges which would eventually lead to an increase in suicides.

Another specific issue requiring attention is the mental health status of disaster workers who were subjected to strong public criticism. A study conducted two-three months after the accident showed that Tokyo Electric Power Company (TEPCO) workers suffered discrimination and slurs, and, as a result, adverse psychological effects. A follow-up study later showed persistent long-lasting effects of discrimination (Hasegawa et al. 2015). Thus, expanded support for affected populations, which takes into account not only dose measurements but also public concerns and priorities related to social and economic upheavals caused by the accident on community well-being and infrastructure, is necessary (Shamisen 2017).

Psychological impacts of previous nuclear accidents

Several decades ago, nuclear accidents had been very rare events. However, it seems that reality is changing with the global expansion of the nuclear industry.

According to energy policy experts and social scientists, there have been more than 100 serious nuclear accidents and incidents from the use of nuclear power (Sovacool 2010). Most notable examples, which resulted in severe environmental contamination and caused serious threats to local communities, include the accidents in Mayak (1957), Sellafield (1957), Three Mile Island (1979), Chernobyl (1986) as well as Fukushima.

Longitudinal epidemiological studies after the Three Mile Island, have reported psychological and emotional concerns about the potential effects of radiation exposure on children, distrust in the radiation experts, and fear of developing cancer among residents (Dew and Bromet, 1993). The scientific results after the Chernobyl have also shown that mental health effects were the most significant public health consequences of the accident and people still suffer from strong concerns about radiation risks (Bromet and Havenaar 2007; UNSCEAR 2008; Tanigawa and Chhem 2014). The Chernobyl accident severely affected three countries and resulted in over 300,000 evacuees. Nowadays, even 30 years after the stress-related symptoms, cognitive and psychological impairments remain among affected people because the indirect health consequences of the affected population, such as psychological issues either have not been addressed or have been addressed very late.

Current psychological and social challenges in Fukushima

Diverse efforts have been made to assess and communicate radiation risks and problems created by the FDNPS accident. Despite the efforts, Japanese society still feels distrust in the government and TEPCO because of their shortcomings in nuclear emergency preparedness and responses, and challenges in the disaster recovery operations (Sato 2016). According to the summary of the Mental Health and Lifestyle Survey conducted by the Fukushima Medical University (FMU), considerable emotional and psychological distress emerged. These are attributable to the unique nature of a nuclear accident and prolonged displacements. As a result, the proportion of those with psychological distress including children is far greater (14.6%) than in other areas affected by the Tohoku earthquake and subsequent tsunami (6.2%) or the Japanese population in other regions (4.2–4.4%). The proportion of residents who currently require support for depressive symptoms and anxieties has been decreasing gradually but remains at significantly higher levels compared with the general population. Some of those mental health issues are directly or indirectly associated with persistent radiation concerns connected to unaddressed risk perceptions and to errors made in risk communication (Maeda and Oe 2017).

Ongoing social and psychological problems among the affected population include:

- risk perception and anxiety regarding radiation;
- psychological and ethical issues related to thyroid ultrasound examinations;
- psychological aspects of nuclear waste disposal;

- discrimination, stigma, self-stigma, and bullying; and
- remaining evacuees, and the dilemma 'to return or not to return'.

Each of these will be addressed in the following sections.

Risk perception and anxiety regarding radiation

Radiation risk perception plays an integral role in shaping individual behaviours for risk accommodation. Factors that govern risk perception or radiation anxiety are different demographic characteristics, disaster-related stressors, trust and radiation-related variables (Takebayashi, Lyamzina, Suzuki and Murakami 2017) as well as the reality of ongoing radiation containment difficulties. It is well-known that members of the public perceive and assess the degree and acceptability of risk differently from radiation protection experts (Slovic 1987; Drottz-Sjoberg and Sjöberg 1991).

Risk perception science consists of many factors which impact anxiety and fear, like a personal stake, control, understanding, dread, and community values (Drottz-Sjöberg and Persson 1993; Oughton 2016; Sjöberg 2000). Many of those risk perception factors have strong psychological, ethical and societal relevance. Therefore, it is a mistake to dismiss public anxiety toward radiation risks as being 'irrational' or 'wrong' (Drottz-Sjöberg and Persson 1993).

The Fukushima experience is a powerful example of the need to effectively address public concerns about the potential risks involved in the accident, recovery and remediation phases, irrespective of whether those concerns are well-founded. Public and stakeholder acceptance of risk, and the efforts to reduce risk, rely not only on understanding the scientific assessment of risk, but also on risk perceptions influenced by several risk perception factors, like dread, volition, familiarity, control, and trust (Slovic 1987; Fischhof 1995; Sjöberg 2000; Ropeik 2002).

Very often several factors operate concurrently. For example, remaining fear of radiation is still high because the nuclear accident was imposed and involuntary. Awareness of the issue remains high and therefore influences one's trust (towards the government and TEPCO), personal stake (lost houses), controllability (issue still remains under control of others, e.g. government) and familiarity.

People tend to ignore information if it does not correspond with their concerns or beliefs and if it is provided through the channels which they do not trust. Furthermore, one of the conclusions of the summary of the Mental Health and Lifestyle Survey in the framework of the Fukushima Health Management Survey is that '*greater perceived radiation risks are associated with poor mental health*'. Also, according to the latest data available from Fukushima, high perceived risk of radiation has a direct correlation with mid-term mental health problems especially among women (Miura et al. 2017). One reason is the fact that, despite all efforts, still too little attention has been paid to the way people perceive and respond to the current risk in and outside of the Fukushima Prefecture.

Most risk communication efforts are currently based purely on knowledge and trust- building campaigns, and, they usually neglect indirect effects like mental health issues or risk perception science. Practice and research show that activities, which are based only on knowledge, are necessary, but insufficient to assuage perceived risk concerns of the affected populations living in different parts of the world (Bromet and Havenaar 2007; Oughton 2016).

Psychological and ethical issues related to thyroid ultrasound examinations

The Chernobyl Nuclear Power Plant accident in 1986, resulted in a marked increase in childhood thyroid cancers mainly due to the intake of radioiodine. Fukushima Medical University (FMU) started the Fukushima Health Management Survey (FHMS) three months after the FDNPS accident in order to monitor the long-term health of residents, promote their well-being, and determine whether long-term, low-dose radiation exposure has any health effects. The FHMS included a thyroid ultrasound examination (TUE) based on the experiences of the Chernobyl accident. The TUE in Fukushima is the world's second largest programme which targeted children under 18 years of age at the time of the accident (initially about 360,000 children).

As of April 2015, about 300,473 (81.7%) children participated in the survey, including elementary and high school students. Among participants, around 2,294 (0.8%) children required confirmatory examination and 116 children had a possible or confirmed malignancy (Fukushima Prefectural Government 2016). In the UNSCEAR Report 2013, it had already been concluded that the occurrence of a large number of radiation-induced thyroid cancers, as observed after the Chernobyl accident, could be discounted because doses were substantially lower (UNSCEAR 2013). However, a higher detection rate of thyroid tumour in a sensitive ultrasound thyroid screening caused not only a great deal of concern among children and parents (Normile 2016; Midorikawa 2016), but also concern among some Japanese academicians' who reported a high prevalence of thyroid tumours in Fukushima Prefecture in 2015 and 2016 (Tsuda et al. 2016; Hiranuma 2016).

In an attempt to address concerns of parents and children about TUE results, FMU started programs aiming to achieve better communication with both children and parents (Midorikawa et al. 2017). First, FMU held explanatory community meetings about the thyroid examination for the parents of children being tested. The meetings took place in smaller facilities, such as schools. Since 2013, over 150 such meetings with more than 8,000 participants have been held.

Second, immediate post-examination individual counselling began in October 2014 – 3 years after the start of the examinations.

Third, in-school class dialogues were initiated in 2015. In that program, FMU teams conducted classes about thyroid examination for students at elementary, junior high, and high schools. The aim was to improve understanding of the meaning of the examination and the interpretation of results. By the end of 2016, about 5,000 children (most of them in grades 5 to 6 of elementary school and junior high school students) took part in the classes. These efforts were effective in facilitating understanding about thyroid examinations and associated diseases. Although TUE is recommended on a voluntary basis for individuals who wish to be monitored, difficulties have been encountered in such a large-scale programme for ensuring properly informed decisions about undergoing thyroid examination by children and parents.

The most recent available data released by the United Nations Committee on the Effects of Atomic Radiation (UNSCEAR) and other sources related to thyroid cancer attribution in Japan challenge the claimed link between radionuclide releases from FDNPS and thyroid cancer increase (Steinhauser, Ortega and Vahlbruch 2017; UNSCEAR 2017). The independent research report from Tsuda et al. (2016) has provoked concern and lends support to UNSCEAR's (2013) 'conservative approach toward continued thyroid surveillance due to the possibility of an increased thyroid cancer risk in the future'. This is why all Fukushima data should be systematically assessed and carefully interpreted for the affected population using standardized criteria, to avoid unnecessary anxiety, confusion, and misunderstanding.

A holistic approach to disaster management and health surveillance is needed. Such an approach will embrace mental health and social and ethical consequences, with the aim '*doing more good than harm*'. It will ensure that assumptions, potential conflicts of interest, risk perceptions and ethical issues and different values will be addressed and, as a result, optimal decision-making for the affected population will be more likely (Shamisen 2017).

Psychological aspects of nuclear waste disposal

It is estimated, that 13–55 per cent of the Fukushima Prefecture will remain above 100mSv 30 years after decontamination activities, and 38–80 per cent of the land will remain above 50mSv (Ministry of the Environment 2017). According to the Ministry of the Environment (2017) of Japan, it is estimated that current amounts of low radioactive waste reached 22 million cubic meters. Current decontamination activities created millions of plastic bags covered with light green plastic, and there now are over 800 so-called 'temporary storage facilities' around the entire Fukushima Prefecture. It is estimated that one day those bags will be moved to the interim storage or final disposal facility depending on the type of waste. Therefore, most probably, current remediation problems will need to be solved by future generations (ICRP Task Group 97 2017).

Discrimination, stigma, self-stigma and bullying

According to the results of the first nationwide survey on bullying, known as 'ijime', conducted between December 2016 and March 2017 by the Ministry of Education, Culture, Sports, Science, and Technology (MEXT) of Japan, 'nearly 200 children evacuated from Fukushima Prefecture following the 2011 nuclear disaster were bullied, with some being abused verbally with derogatory terms linked to the calamity' (Aoki 2017). Cases were clearly linked to the nuclear accident and took place in elementary schools, with evacuees being told to 'go back to Fukushima', being called 'germ' and 'radioactive' by classmates, and being blamed for the nuclear power plant explosions. Bullying is dangerous because it always has harmful and destructive effects on mental health and self-perception, especially among youth and teenagers. This is why discrimination against Fukushima evacuees and Fukushima children is a serious problem and, if not addressed, anxieties will stay with them for many years.

Remaining evacuees, and the dilemma 'to return or not to return'

Along with progress in reopening affected municipalities included in the official evacuation zone, evacuees have been encountering a dilemma on whether or not to return, or where else they should settle if not to return. Evacuees are put in the difficult position of having to make a decision, one that would largely influence their future and that of their family. And, these decisions are made under increasing social pressure, such as from the government through the gradual termination of financial aid and other assistance (Sato 2017).

Conclusion

Six years have passed since the March 2011 accident occurred in Japan. Despite the initial confusion, actual dose measurements across Fukushima Prefecture are now well under control. Internal radiation exposures from radiocaesium are not significant and external exposure is not very different or even lower than those in other parts of Japan or other countries like France, Poland or Belarus. A large amount of empirical data has been collected by different stakeholders and have been disseminated through different channels (Hayano 2016; Safecast 2017). Unfortunately, however, awareness of available results remains poor within Japan and worldwide (Hayano 2016).

Despite the fact that radiation levels have significantly declined (UNSCEAR 2017), the nuclear accident continues to affect the life of people in Fukushima Prefecture because residents continue facing diverse social and psychological issues. Psychic and mental health consequences, anxiety, stress and remaining high perception of radiation risk along with socio-economic challenges are considerable. And, despite visible progress in decontamination and remediation activities, the clean-up process is expected to continue for the next 30–40 years. The people of Fukushima thus will be facing long-term challenges well into the future.

The Fukushima accident has revealed that it is vital to monitor both direct (i.e., more immediate effects) and indirect (i.e., other health and psychosocial problems) health consequences arising from a nuclear accident and to pay special attention to the complex psycho-social health effects which they cause. The Fukushima experience shows that it is not sufficient to focus only on technical aspects of radiation protection because indirect consequences, such as 'psycho-social issues', play the major role in how communities and affected populations cope with disasters (Oughton 2016). This is why there is an urgent need to carefully identify, analyse and understand prevalent perceptions among affected communities, especially among socially vulnerable groups (e.g. mothers with small children, elderly, and evacuees), and to develop a systematic approach for psycho-social factors among communities living in the territories affected by nuclear accidents.

For successful risk communication, there is a need for multidisciplinary and holistic approaches to disaster management into which would be integrated: social, psychological, economic, ethical, radiation protection and health surveillance issues. The aim is to achieve more good than harm (Shamisen 2017) and lead to an increase in trust and improvements in the well-being of affected communities. The tailoring of risk messages must be responsive to perceptions of the different demographic groups residing within affected communities.

Acknowledgements

The views expressed in this article are solely those of the author and do not necessarily reflect the views of the Fukushima Medical University or any other authority. I thank Ms Akiko Sato, Dr Koichi Tanigawa and Dr Ken Nollet for their support and valuable contributions to this manuscript. I am especially grateful for the valuable advice, help and support of Dr. Colin L. Soskolne in the writing of this chapter.

References

Aoki, M. (2017) Survey Finds Bullying against Young Fukushima Evacuees in Schools. *Japan Times*, 11 April.

Becker, S. M. (2013) Psychological Issues in a Radiological or Nuclear Attack. In United States Government US Army (eds), *Medical Consequences of Radiological and Nuclear Weapons*. Washington, DC: United States Government.

Bromet, E. J. and Havenaar, J. M. (2007) Psychological and Perceived Health Effects of the Chernobyl Disaster: A 20-year Review. *Health Phys* 93: 516–521.

Dew, M. and Bromet, E. J. (1993) Predictors of Temporal Patterns of Psychiatric Distress during 10 Years Following the Nuclear Accident at Three Mile Island. *Soc Psychiatr Epidemiology* 28: 49–55.

Drottz-Sjöberg, B.-M. and Persson, L. (1993) Public Reaction to Radiation: Fear, Anxiety, or Phobia? *Health Physics* 64: 223–231.

Drottz-Sjöberg, B.-M. and Sjöberg, L. (1991) Adolescents' Attitudes to Nuclear Power and radioactive wastes. *Journal of Applied Social Psychology* 21: 2007–2036.

Fischhof, B. (1995) 'Risk Perception and Communication Unplugged: Twenty Years of Process. *Risk Analysis* 15(2): 137–145.

Fukushima Prefectural Government (2016) *Report of the Fukushima Health Management Survey*. Fukushima: Fukushima Prefectural Government.

Fukushima Prefectural Government (2017) *Steps for Revitalization in Fukushima*. Fukushima: Fukushima Prefectural Government.

Hasegawa, R. (2013) Disaster Evacuation from Japan's 2011 Tsunami: Disaster and the Fukushima Nuclear Accident. *Climate* 5 (13 May).

Hasegawa, A., Tanigawa, K., Ohtsuru, A. (2015) Health Effects of Radiation and Other Health Problems in the Aftermath of Nuclear Accidents, with Emphasis on Fukushima. *Lancet* 386(9992): 479–488.

Hayano, R. (2016) Measurement and Communication: What Worked and What Did Not Work in Fukushima. *In ICRP, Proceedings of the International Workshop on the Fukushima Dialogue Initiative*. Ann. ICRP 45(2S). Ottawa: International Commission on Radiological Protection.

Hiranuma, Y. (2016) *Fukushima Thyroid Examination Fact Sheet*. Version 2e. Fukushima: Fukushima Voice.

IAEA (2015) *The Fukushima Daiichi Accident*. Report by the Director-General. Vienna: IAEA.

ICRP Task Group 97 (2017) Meeting of the Group Related to Application of the Commission's Recommendations for Surface and Near Surface Disposal of Solid Radioactive Waste. 5–10 November, Fukushima City, Japan.

Kingston, J. (2012) *Natural Disaster and Nuclear Crisis in Japan. Response and Recovery after Japan's 3/11*. Nissan Institute/Routledge Japanese Studies Series. Abingdon: Routledge.

Maeda, M. and Oe, M. (2017) Mental health Consequences and Social Issues After the Fukushima Disaster. *Asia Pac J Public Health* 29(2S): 36S–46S.

Midorikawa, S., Suzuki, S. and Ohtsuru, A. (2016) After Fukushima: Addressing Anxiety. *Science* 352(6286): 666–667.

Midorikawa, S., Tanigawa, K., Suzuki, S. and Ohtsuru A. (2017) Psychosocial Issues Related to Thyroid Examination after a Radiation Disaster. *Asia Pac J Public Health* 2(suppl): 63S–73S.

Ministry of the Environment (2017) Environmental Remediation. Available at http://josen.env.go.jp/en/decontamination.

Miura, I., Nagai, M., Maeda, M. (2017) Perception of Radiation Risk as a Predictor of Mid-Term Mental Health after a Nuclear Disaster: The Fukushima Health Management Survey. *Int J Environ Res Public Health* 14(9).

Morino, Y., Ohara T. and Nishizawa, M. (2011) Atmospheric Behavior, Deposition, and Budget of Radioactive Materials from the Fukushima Daiichi Nuclear Power Plant in March 2011. *Geophys. Res. Lett.* 38: L00G11.

Normile, D. (2016) Epidemic of Fear. *Science* 351(6277): 1022–1023.

Oughton, D. (2016) Societal and Ethical Aspects of Radiation Risk Perception. In J. Shigemura and R. Chhem (eds), *Mental health and Social Issues Following a Nuclear Accident*. Berlin: Springer.

Ropeik, D. (2002) *Understanding Factors of Risk Perception*. Cambridge, MA: Nieman Reports.

Safecast (2017) *Report Part 2.1: Situation Report – Issues at Fukushima Daiichi Nuclear Power plant (FDNPP)*. Tokyo: Safecast.

Sato, A. (2016) Nuclear Disasters, and Risk Communication: Learning from Fukushima. *Policy Brief* 5.

Sato, A. (2017) Information Needs and Modalities among People Affected by the Fukushima Nuclear Disaster. *Universal Journal of Management* 5: 67–79.

Shamisen (2017) *Recommendations and Procedures for Preparedness and Health Surveillance of Populations Affected by a Radiation Accident*. Shamisen.

Sjoberg, L. (2000) Factors in Risk Perceptions. *Risk Analysis* 20(1): 1–12.

Slovic, P. (1987) Perception of Risk. *Science* 236: 280–286.

Sovacool, B. (2010) *Journal of Contemporary Asia* 40(3): 393–400.

Steinhauser, G., Brandl, A. and Johnson, T. E. (2014) Comparison of the Chernobyl and Fukushima Nuclear Accidents: A Review of the Environmental Impacts. *Sci Total Environ* 470–471: 800–817.

Steinhauser, G. and Saey, P. R. J. (2016) 137Cs in the Meat of Wild Boars: A Comparison of the Impacts of Chernobyl and Fukushima. *J Radioanal Nucl Chem* 307: 1801.

Steinhauser, G., Ortega, M. C. and Vahlbruch, J-W. (2017) Japanese Food Data Challenge the Claimed Link between Fukushima's Releases and Recently Observed Thyroid Cancer Increase in Japan. *Scientific Reports* 7: 10722.

Takebayashi, Y., Lyamzina, Y., Suzuki, Y. and Murakami, M. (2017) Risk Perception and Anxiety regarding Radiation after the 2011 Fukushima Nuclear Power Station Accident: a Systematic Qualitative Review. *Int J Environ Res Public Health* 14(11): E1306.

Tanigawa, K. and Chhem, R. (2014) *Radiation Disaster Medicine: Perspective from the Fukushima Nuclear Accident*. Berlin: Springer.

Tsuda, T., Tokinobu, A., Yamamoto, E. and Suzuki, E. (2016) Thyroid Cancer Detection by Ultrasound Among Residents Ages 18 Years and Younger in Fukushima, Japan: 2011 to 2014. *Epidemiology* 27(3): 316–322.

UNSCEAR (2008): *Health Effects Due to Radiation from the Chernobyl Accident*. Vienna: UNSCEAR.

UNSCEAR (2013) *Levels and Effects of Radiation Exposure due to the Nuclear Accident after the 2011 Great East-Japan Earthquake and Tsunami*. Vienna: UNSCEAR.

UNSCEAR (2017) *Developments since the 2013 UNSCEAR Report on the Levels and Effects of Radiation due to the Nuclear Accident following the Great East-Japan Earthquake and Tsunami. A 2017 White Paper to Guide the Scientific Committee's Future programme of work*. Vienna: UNSCEAR.

11 Civil society preventing environmental disasters

Anne Venton

Introduction

This chapter is a case study of the role of civil society in advocating for the interest of the broader public in nuclear safety. The case is the Ontario Power Generation's (OPG) proposal to construct a Deep Geologic Repository (DGR) for burying low and intermediate level nuclear waste (L&ILW) very near the shores of Lake Huron at Kincardine, Ontario not far from the Bruce Nuclear Generating Station. More specifically, civil society has been working for over a decade in demanding safety standards and protocols to prevent potential nuclear disasters that could result from this project. It is argued that the Ontario government should put public safety first by making Ontario's nuclear emergency plans the best possible. However, at present, Greenpeace reports that 'Ontario is not prepared for a nuclear emergency' (Stensil 2017b).

The role of civil society in environmental stewardship

Civil Society is the aggregate of non-governmental organizations and institutions that manifest the interests and the will of citizens. It is often referred to as the third sector of society, distinctly different from the business and government sectors. Its function is to correct the imbalance in power and influence on government between the business sector and the general public which is often uninterested or uniformed about government decisions that significantly affect their lives. This imbalance is characteristic of representative democracies as opposed to direct democracies.

In her 2014 book, *This Changes Everything: Capitalism Versus the Climate*, Naomi Klein argues that the environmental movement is now over 50 years old and has made little progress in that time. Since 1990, when negotiations toward a climate treaty began in earnest, global carbon dioxide emissions have increased 61 percent (Klein 2014: 11). To date, progress in averting the climate crisis has been too slow and too inadequate. The reason for this lack of progress is that the proponents in the environmental movement for averting the crisis of climate warming have been strongly opposed by the proponents of the 'reigning economic paradigm of deregulated capitalism and public austerity' (ibid.: 63). And 'right now capitalism is winning (the battle) hands down' (ibid.: 22). In the context of neo-liberal capitalism the strategies and tactics of the opponents of the environmental movement have included the following (ibid.: 38):

- Oil and gas companies have denied that climate change is caused by their companies.
- Capitalists have funded think tanks to change public attitudes to allow reduced regulation on companies generating greenhouse gases.
- Relentless corporate advertising in the media has extolled material consumption as a virtue.

The 'Blockadia' movement

In response to the duplicitous strategies of the business community to slow progress in protecting the environment and addressing climate change, the environmental movement has matured over the last decade and with it, civil society organizations have expanded. All of this is described in a chapter of Klein's book entitled 'Blockadia: The New Climate Warriors'. Blockadia is the name of a grassroots movement of civil societies for the purpose of preserving the natural environment and addressing climate change. Its tactics include civil disobedience activities such as blocking pipelines and blocking fracking processes – hence the name Blockadia. It has expanded the movement by uniting indigenous and other activist groups in these endeavours.

Its members challenge corporate marketing with academic research that identifies the sacrifices foisted on people by oil and gas businesses. On behalf of those being sacrificed, its members put pressure on community political leaders to stop oil and gas development.

The movement initiates mass protests like the civil rights movement to slow down the economic machine for environmental destruction (Klein 2014: 450). Foremost among the mass movements were the 40,000 people outside the White House on February 2013 protesting the proposed Keystone XL pipeline (ibid.: 302). The movement is also positioned as a mass global movement as evidenced, for example, by the 'Global Frackdown' in September 2012, with actions in two hundred communities in more than twenty countries (ibid.: 304).

It is a pro-democracy movement that insists on the rights of people in communities to have a say in critical decisions relating to water, land and air. For example, it was instrumental in helping 170 New York towns adopt anti-fracking ordinances (ibid.: 361).

The movement is also a constructive one that advocates building alternative institutions to replace some of the large capitalist firms in the economy. These alternatives include cooperatives in which profits are shared with stakeholders, and non-profit organizations with missions to serve community interests rather than financial returns on equity. The alternatives also include local communities that can feed themselves thereby being less vulnerable to price shocks within the broader globalized food system. That is why La Via Campesina, a global network of small farmers with 200 million members, often declares 'Agroecology is the solution to solve the climate crises' (ibid.: 134).

Finally, Blockadia aims to correct the imbalance of power between the capitalists and the environmentalists in civil society in two ways. First, it aims to turn the tables by insisting that it is up to industry to prove that its methods are safe rather than the people who will be harmed by them (ibid.: 335). Second, it challenges the decades-long practice of risk assessment in which environmentalists worked with partners in business and government to try to balance the risks of dangerous levels of pollution against the need for profit and growth in GDP in order to find acceptable levels of risk. Against this, Klein argues that, when human health and the environment are significantly at risk in the sense that science cannot confirm that the risks are insignificant, the precautionary principle requires that no action should be taken. And so she concludes, '[in] Blockadia, risk assessment has been abandoned on the barricaded roadside, replaced by a resurgence of the precautionary principle' (ibid.: 335).

According to the Canadian Environmental Law Association (CELA), the precautionary principle denotes a duty to prevent harm when it is within our power to do so and even when all the evidence is not in. This principle has been codified in several international treaties to which Canada is a signatory (CELA). Klein discusses the precautionary principle in the context of environmental reform. She argues that when human health and the environment are significantly at risk in the sense that science cannot confirm that risks are insignificant, no action should be taken. As well, she argues that the burden of proof that a practice is safe should not be placed on the element of the public that is being harmed.

Naomi Klein's 'focus in *No Is Not Enough* (2017) remains the vociferous advocacy for grassroots, popular political movements that can meaningfully push back against [Donald Trump, and] the governing political-economic logic of neoliberalism more generally' (Semley 2017: R15).

History and conceptualization of the DGR project

The DGR project is a proposal by the OPG for site preparation, construction, operation, decommissioning and abandonment of a deep geologic waste disposal facility for low and intermediate level radioactive waste. It is proposed that the facility be located near the Bruce nuclear generating station adjacent to OPG's Western Waste Management Facility (WWMF) within the municipality of Kincardine, Ontario. The facility would manage L&ILW from OPG owned nuclear generating stations at Pickering and Darlington nuclear sites in Ontario as well as the Bruce nuclear generating station. The proposal is divided into two phases. First is the pre-closure phase including a 60 year period encompassing site preparation, construction, operating and decommissioning. The second is the post-closure phase including institutional control followed by abandonment forever (Ontario Power Generation 2007).

Dangers of the proximity of the DGR to Lake Huron

Civil society groups have expressed concerns about the location of the proposed DGR approximately 1.2 km from the shore because they fear radioactive contamination of the Great Lakes. The site of the proposed DGR, to be constructed in limestone, is adjacent to Lake Huron near the Bruce Nuclear Generating station, the largest in the world. Lake Huron is one of the five interconnected Great Lakes of North America which contain 21 per cent of the world's surface fresh water covering 94,250 square miles which provides drinking water to over 30 million residents of Canada and the United States who live in the Great Lakes basin. This area holds eight of Canada's 20 largest cities: Toronto, Hamilton, London, St Catharines, Niagara Falls, Oshawa, Windsor and Barrie. As well, the US cities of Duluth, Milwaukee, Chicago, Gary, Detroit, Toledo, Cleveland, Erie, Buffalo and Rochester, are located on the shores of the Great Lakes.

OPG's proposal also involves transportation of hazardous nuclear waste to the DGR from other nuclear generating stations located in Ontario including Darlington and Pickering. It is proposed that L&ILW will be transported by truck, rail or ship to the Bruce site through heavily-populated areas. Two million people live within 30 kilometres of the Pickering Nuclear Station, more than any other nuclear plant in North America. As well, Pickering is the fourth oldest nuclear plant in North America and the seventh oldest in the world. OPG is seeking permission to continue to operate Pickering's aging nuclear reactors until 2024 which is 6 years beyond its operating licence which expires on August 31 2018. Refurbishing the reactors will be expensive and the transport of waste dangerous. Critics have suggested closing down the Pickering and Darlington reactors and instead purchasing hydroelectric power generated in Quebec which would be less expensive and less hazardous to transport to Ontario (see www.ontariocleanairalliance.org).

Government policy and legislation updates

Since 1974, L&ILW from Bruce, Pickering and Darlington reactors has been stored centrally on the Bruce nuclear site in engineered storage structures. However, new government policies and legislation have prompted the OPG's proposed DGR project. In 1996, the Government of Canada's Radioactive Waste Policy Framework (RWPF) tasked OPG to implement a long term solution for managing its L&ILW. The Nuclear Fuel Waste Act (NFWA) legislation came in to effect Nov. 15 2002 which dealt with management of used nuclear fuel.

In 2004, the Municipality of Kincardine, Ontario located on Lake Huron signed a Willing Host Memorandum of Understanding to set out terms under which OPG, in consultation with Kincardine, would develop a long-term management plan for dealing with L&ILW at the WWMF. A Report was published in 2004 which included technical feasibility and socio-economic impacts resulting in the preferred option of a DGR constructed in limestone which was adopted by the Municipality of Kincardine on 21 April 2004. In the terms of the Willing Host agreement, Kincardine and adjacent municipalities have received not only millions of dollars for over a decade but also the expectation of new jobs in the area, a vocational school and a centre of energy excellence. Further approvals for the proposed DGR were required by the Canadian Nuclear Safety Commission (CNSC) and a federal environmental assessment under the Canadian Environmental Assessment Act (CEAA 1999). In June 2007 the environmental assessment of the proposed DGR Project was referred to a Joint Review Panel which included a 75-day public comment period. An Environmental Impact Statement (EIS) followed and OPG submitted its EIS in 2011. When the updated CEAA (2012) came into effect, officials deemed that the project review was to continue under the new legislation. As a result, the panel encouraged participation of government agencies, members of the public, and Aboriginal groups throughout the process. Public hearings were held in Kincardine, Saugeen Shores and on-line through the Canadian Environment Assessment Registry (CEAR). Funding was provided for participants. On 6 May 2015, the Panel Report endorsed the DGR project proposal (Wikipedia 2017).

Canadian Federal and Provincial Government Input

Both the Canadian federal government and the Ontario provincial government have input regarding the construction and implementation of OPG's proposed DGR and have been calling for input from citizen groups regarding this Proposal. In early 2016, the current federal Minister of Environment and Climate Change in an open letter to interested parties delayed the decision on whether to approve OPG's proposed DGR for L&ILW and requested a short- term deadline for the OPG to furnish a time frame within which it could provide an updated list of commitments to mitigate potential damage from the site. The Minister of Environment and Climate Change is the Minister of the Crown in the Canadian Federal Cabinet who is responsible for overseeing the policy and actions of this department.

In February 2016, the Minister of Environment and Climate Change requested additional information and further studies from OPG on the environmental assessment for the proposed DGR. These additional requirements extend the date of the decision on the project proposal into late 2017 or beyond. The additional information request included:

- alternate locations for the project;
- cumulative environmental effects of the project; and
- an updated list of mitigation commitments for each identified adverse effect under the CEAA 2012.

Opposition from civil society in Canada and the US

Many communities in the Great Lakes basin have passed resolutions against OPG's DGR project. More than 160 cities, counties and states around the Great Lakes oppose the DGR including Ontario cities of Toronto, Mississauga, Hamilton, Niagara Falls, Kingston, Thunder Bay, Sault St. Marie, Windsor and more, as well as local governments in the states of Michigan, Illinois, Pennsylvania, Indiana, Minnesota, Wisconsin, New York and Ohio. For example, the Michigan Senate prohibits waste burial within 16 kilometres of the shore. Another major concern of the above communities is the lack of internationalization of management processes. The current process limits discussion to local level consultation.

In 2013 Michigan's Senate adopted resolutions to urge Canadian officials to thoroughly review OPG's DGR proposal. The Michigan Senate also asked the US President and Secretary of State and the US Congress to intervene, and called for increased involvement of the International Joint Commission (IJC), the Great Lakes Commission (GLC) and all Great Lake States as well as Ontario and Quebec. The resolution also quotes Michigan's criteria which states the storing of low-level nuclear waste is prohibited within ten miles (16 km) of Lake Michigan, Lake Superior, the St. Mary's River, Lake Huron, Lake Erie, the Detroit River, the St. Clair River or Lake St Clair. Opposition from Michigan's two US senators and members of the US House of Representatives continues. On 26 September 2015, another resolution was passed opposing the proposed DGR by the Great Lakes Legislative Caucus (GLLC). This is an external group of state and provincial lawmakers from eight US states and two Canadian provinces (Ontario and Quebec).

Condemnation of OPG's proposed DGR project by NGOs and grassroots groups

Spokespersons for Citizens' Groups in both the US and Canada expressed concerns regarding health and procedural issues relating to OPG's current Report on the Northwatch webinars on 14 and 21 February 2017, which I attended. The executive director of Northwatch acted as moderator of these two webinars which were part of a larger series which included representatives of community groups and NGOs in both Canada and the US. Key issues cited by participants in the webinars included concern that the burial of nuclear waste in limestone was unprecedented and, as a result, there are many unknowns regarding complications of L&ILW to be stored in the DGR site. Another concern was the OPG's failure to answer questions and provide details about alternative locations for the DGR. Still another concern was that the OPG would not release cost estimates in its latest report. It was suggested that OPG information was out of date and ill-informed. Northwatch prepared an extensive report on what it felt were the weaknesses of the OPG Report that was posted on the internet (Wikipedia 2017).

Citizen groups in Canada and the US echo Northwatch's concerns

Other groups echoed the concerns expressed in the Northwatch webinars – concerns about the lack of precedents and the limited scope of testing found in OPG's reports. NGOs such as Stop the Great Lakes Nuclear Dump Group (SGLNDG) condemned the OPG reports and drew up a petition with 62,000 names. Grass roots organizations established in 2012 included: Save Our Saugeen Shores later renamed Save Our Shores Great Lakes (SOSGL); The United Tribes of Michigan, representing 12 First Nations and other indigenous groups; and First Nations and Metis organizations in Canada who opposed the DGR proposal. For example, Saugeen First Nation voiced opposition in May 2015. NGOs and Community groups are important because they provided an independent point of view of the OPG report.

Accident risks and long-term safety

Major concerns voiced by citizen groups were lack of research and and absence of precedents for constructing a DGR in limestone. Since 1978, the Canadian Nuclear Safety Commission (CNSC) had explored options for safely storing used nuclear waste in granite. However, CNSC has only been researching sedimentary rock as a suitable location for safely storing used nuclear waste since 2014. To date, DGRs in Europe, the US and Asia have been constructed in granite, tonalite, clay and salt domes. Although there are multiple operational DGR sites worldwide, only those constructed in granite geologic formations have operated without incident.

Professor Erika Simpson notes that the scientific and technical merits and demerits of the DGR are already well documented in various reports and hearings (Simpson 2016). But more controversy is expected now that the current federal Minister of the Environment and Climate Change asked for more information about the project in February 2016 (Simpson 2017). As mentioned earlier, requests for alternate locations for the project, for the cumulative effects of the project and for an updated list of mitigation commitments for each identified adverse effect under the CEAA (2012) were made.

Some critics argue that a DGR for L&ILW should not be constructed so close as 1.2 kilometres from Lake Huron. Michigan's criteria states that the storing of low-level nuclear waste is prohibited within ten miles (16 km) of Lake Michigan, Lake Superior, the St. Mary's River, Lake Huron, Lake Erie, the Detroit River, the St. Clair River or Lake St. Clair. Professor Simpson notes that 'the only example that OPG offered in its Report of a similar DGR was the Waste Isolation Pilot Plant in New Mexico. But that facility is no longer operating, after an underground fire and loss of containment resulted in radioactive releases to the surface in 2014' (Simpson 2017). Research conducted in China has shown that exposure to mechanical and chemical weathering has been proven to mechanically destabilize limestone, leading to geo-hazards that include sink holes and rapid erosion of limestone into soil. This could possibly occur during construction of the DGR and during its highly extended lifetime.

Safety standards and protocols for potential nuclear disasters

Greenpeace's Shawn-Patrick Stensil reports that Ontario has only detailed plans for relatively small accidental radiation events. Given its plans to refurbish aging reactors and proposed plans for the construction for a DGR to store L&ILW, Ontario needs detailed plans to protect vulnerable communities and drinking water as well as ensuring transparency and public participation which meet or exceed international best practices. As well, government should require regular five-year reviews and detailed consultations with the public and affected communities to ensure continuous improvement of both the planning basis and emergency response measures (Stensil 2017a).

In 2013, the Ontario Minister of Community Safety made several commitments to civil society organizations related to public engagement and addressing gaps in Ontario's offsite nuclear emergency plans. In 2017, the government released a discussion paper on how it plans to update its nuclear emergency plans in response to the Fukushima accident. Greenpeace has issued a Report Card stating that the Ontario government should publish modelling on the impacts of a Fukushima-scale accident before it provides final approval for its plan to continue operating the Pickering reactor post 2018. As well, there is no evidence that the government has launched a campaign to inform citizens of the availability of potassium iodide pills within the 50 km Secondary Zone as well as providing a detailed evacuation plan, especially for vulnerable communities (Stensil 2017c).

Greenpeace's Report Card on Ontario's Nuclear Energy Review was very negative. It gave the Province a failing grade regarding the protection of drinking water, land recovery and evacuee return. It noted that the Province's commitment to begin a process to achieve greater transparency in public engagement relating to nuclear-related matters had not happened. Further, it criticized the government because it continued to consult with industry behind closed doors without public oversight and because it had systematically refused to respond to Freedom of Information requests (Stensil 2017c).

Conclusions

Civil society has concluded that the proponents of the DGR project ignored a significant possibility of a nuclear disaster that could affect millions of people living in the Great Lakes basin as well as people living in communities in Ontario and bordering American states through which nuclear waste is to be transported by road, train and boat to the DGR. In short, civil society has concluded that the DGR proposal ignores the precautionary principle.

This case study also shows how civil society has followed Naomi Klein's dictum that the burden of proof should be borne by the proponents of the DGR rather than the general public who will be subjected to the risk of nuclear disasters. Civil society has helped to identify the failure of the proponents to provide adequate proof. More importantly the result has been a far greater dissemination of information and publicity about the possible impacts of the project. For, example a *Globe and Mail* article noted that the new OPG Report 'did little to assuage [the] concerns about the potential for catastrophic pollution of a major drinking water source voiced by scores of groups and communities in both Canada and the US' (Perkel 2017). Media publicity can effect decisions of government. Hopefully this will tip the balance of influence on government policy decisions between the highly organized and funded business community and the largely unorganized democratic majority in favour of the latter. The government should listen to and incorporate civil society proposals to tackle the problems of possible nuclear environmental disasters.

References

Klein, Naomi (2014) *This Changes Everything: Capitalism Versus the Climate*. Alfred A. Knopf, Toronto, Canada.

Klein, Naomi (2017) *No Is Not Enough: Resisting Trump's Shock Politics and Winning the World We Need*. Alfred A. Knopf, Toronto, Canada.

Ontario Power Generation (2007) 'Overview of OPG's Proposed L&ILW Deep Geologic Repository Bruce Site, Tiverton, Ontario', www.opg.com/generating-power/nuclear/nuclear-waste-management/Deep-Geologic-Repository/Documents/HPD/4.3.53_overviewofontariopowergene.pdf (accessed 2 June 2017).

Perkel, Colin (2017) 'Lake Huron Site Best Fit for Nuclear-Waste Dump', *The Globe and Mail* 31 May, pA9.

Semley, John (2017) 'Surviving Trump: Naomi Klein's Latest Book is a Cautiously Hopeful Document for a Despairing Age', *The Globe and Mail* 17 June, pR15.

Simpson, Erika (2017) 'The Federal Government Wants to Know So Much More about a Proposed Nuclear Waste Site Near Lake Huron, It Seems It Will Never Get the OK', *Postmedia Network* 21 April.

Simpson, Erika (2016) 'Nuclear Waste Burial in Canada: The Political Controversy over the Proposal to Construct a Deep Geologic Repository' *Journal of Nuclear Energy Science and Power Generation Technology*, vol. 55, no. 3, April, pp. 1–6.

Stensil, Shawn-Patrick (2017a) 'A Call for Public Safety: Addressing Nuclear Risks on the Great Lakes', Greenpeace.org (accessed 1 August 2017).

Stensil, Shawn-Patrick (2017b) 'Key Messages: Speak Out for Public Safety', Greenpeace.org (accessed 1 August 2017).

Stensil, Shawn-Patrick (2017c) 'Ontario's Nuclear Emergency Review: Report Card', Greenpeace.org (accessed 1 August 2017).

Wikipedia (2017) 'Deep Geologic Repository', www. wiki2.org/en/Ontario_Power_Generation (accessed 3 June 2017).

12 A global update on the ambit of unconventional gas mining and an alternative framework for mediating energy demands

Janice Gray

Introduction

The exploitation of large deposits of unconventional gas (UG) has been hailed as a way out of the so-called looming 'energy crisis' that arguably threatens the regular supply of energy to households, businesses and other entities throughout the world. Accordingly, the international community has witnessed a widespread, rapid but uneven uptake of unconventional gas mining (UGM) and the associated techniques of hydraulic fracturing (fracking) and horizontal drilling.

This chapter offers a global survey and update of UGM activities, one purpose of which is to highlight the *reach* of UGM activities and underscore the potential magnitude of harm that could ensue if UGM is not well governed. The chapter identifies jurisdictions from Africa to Europe, Australia and the Americas as being sites of variously advanced levels of UGM activities. It also identifies jurisdictions where UGM and fracking have been prohibited to demonstrate the degree of concern and anxiety about the industry that simultaneously exists alongside expansionist tendencies. Given that UGM is a dynamic field such an update is considered important.

The potential problems and negative outcomes associated with UGM and fracking are well rehearsed in other literature and are not discussed in detail here. They include: aquifer contamination, fugitive gas emissions, produced water (Herlihy 2009: 461), flaring incidents, well (dis)integrity, gas capture and flaring, seismicity, well remediation on termination of activities, human and animal health, community engagement tensions between competing future land uses, pressure on landowners to engage with resource companies resulting in time loss and the diminishing productivity of their own businesses (GISERA 2014; Meng 2016; Gray and McCoy 2016; Australian Senate 2016; Grafton et al. 2017). Jurisdictions have differed on which governance tools are most effective to manage these problems and outcomes. Arguably the success of all tools will be challenged while consumption tends to dominate lifestyles.

Accordingly, another objective of this update is to reinforce the extent of global progress towards planetary limits and ecological tipping points (Rockström et al. 2009). Factors such as consumerist lifestyles along with unpredictable weather conditions, population growth, and the oversupply of generation all place high demands on the energy sector as the state-wide power blackout in 2016 in South Australia demonstrated. Such blackouts tend to result in calls for additional energy sources, such as UG, rather than a reassessment of unsustainable ways of living.

In their encouragement of the UG industry, many political leaders have rejected much of the criticism associated with UGM, arguing that effective regulation, particularly monitoring and compliance *will* maintain rigorous health, safety and environmental standards but regulation is unlikely to succeed without adequate funding and support (see Willacy 2017). More generally expansion of UGM seems short sighted and likely to lead to citizen and environmental vulnerability. It is an approach that is ultimately unsustainable and lacks ecological integrity. Why? One obvious reason is that UG is simply not a renewable energy source. It is a fossil fuel and although cleaner than coal, still generates greenhouse gas emissions (Brandt et al. 2014). Responding to energy shortages by depleting more and more non-renewable natural resources (to generate energy) represents further anthropogenic stress on the environment. It has the effect of denying land and waterscapes their biological potential and takes the planet closer to its ecological limits. Some climate change experts are worried that even if UG is only used as a transitional fuel it may prevent developed nations from meeting their Paris carbon reduction targets (La Fleur and Sandford 2016). It may sound in irretrievable planetary damage. Yet, a pattern of UG uptake continues across the globe as the material below demonstrates.

Unconventional gas mining: a global survey and update

Recoverable UG resources (that is, natural gas resources that are exploited by unconventional methods) are mainly concentrated in 37 countries including the United States (US), China, Russia, Canada, Australia, Iran, Saudi Arabia, Argentina, Libya and Brazil (Wang et al. 2016: 931). Although parts of Australia are aggressively pursuing UGM, the United States (US) remains the frontrunner in terms of production. In other jurisdictions exploration only has commenced while in others still, there is little or no will to embrace UGM at all. The different rates and extent of UGM have been tied to 'differing geological conditions, the lack of skilled and experienced human and capital resources of established markets or the result of unpredictable above-ground conditions such as taxation, foreign investment and intellectual property issues' (Grafton et al. 2017: xiv). They are also linked to different political and socio-cultural climates.

US

UGM began in Texas (still the largest shale gas supplier in the US) but spread to the Marcellus Basin (in Pennsylvania, West Virginia and Ohio) where uptake has been rapid. In 2017, one fifth of US gas production came from the Marcellus region (Grafton et al. 2017: 2). Shale gas is mined in at least 16 states including, Wyoming, Louisiana and Oklahoma (EIA 2016). Overall shale gas reserves in the US are estimated to be 623 trillion cubic feet or 17.6 cubic metres (World Energy Council 2016: 15).

Coal bed methane mining (CBMM) is less extensive but occurs in Colorado, Wyoming and New Mexico, for example. Fracking regulation, in particular 'is minimal, especially at the federal level as a result of the Energy Policy Act of 2005 and the "Halliburton Loophole"… The Halliburton Loophole exempts oil and gas companies from key environmental protection laws such as the Clean Water Act and the Safe Drinking Water Act, among others' (Doyon and Bradshaw 2015: 235).

Canada

In Canada tight gas production (rather than shale gas or CBM) has dominated UG production from the mid-1980s (Underschultz 2017: 551 relying on Masters 1984).

Nevertheless, CBM has been produced in Canada in small volumes since about 2000 with production escalating in the mid-2000s. Shale gas production did not really take off until 2009 and has remained behind US developments (Underschultz 2017: 545, 551–2). There is known shale gas resource potential in at least five areas of Alberta (Natural Resources Canada 2017: fig. 1).

UGM is regulated on a province by province basis and current, national information especially on fracking sites is difficult to assemble.

South Africa

It is estimated that the Karoo is the eighth largest shale reserve in the world with up to 390 million cubic feet of technically recoverable shale reserves (Corrigan and Murtazashvili 2016). In 2016, a government commissioned report noted that (a) exploration right applications had been accepted for shale gas in the Karoo but were still under consideration and (b) that government would 'seriously consider' shale gas development on a significant scale if found to be economically viable (Scholes et al. 2016: preface and ch. 1).

Twenty five exploration rights for CBM have been awarded and some companies have applied for production rights. Five technical co-operation permits have been issued for shale gas exploration (SAOGA undated).

China

In 2011, Chinese CBM production grew more than 20 per cent (World Energy Council 2016: 19). It is anticipated that production will be at 14 billion cubic metres (bcm) by 2020 (ibid.: 20).

Meanwhile shale gas production was 1.3 bcm in 2014 and although China has plans to replicate US shale gas activities, the Sichuan Basin, where shale gas resources are located, is a highly populated, mountainous region lacking in the large volumes of water necessary for UGM. Nevertheless, by early 2015 the

Sichuan Basin had reached 4.4 bcm of production and had production rates of 60,000 to 200,000 cubic metres per day per well which are comparable to production in the prolific Marcellus basin in the US (ibid.: 20).

Saudi Arabia

According to the World Energy Council (2016: 22), 'Saudi Arabia is set to become a shale gas producer by 2020 and the nation's NOC, Saudi Aramco, recently announced it is investing another $7 [billion] to develop shale gas resources.'

Mexico

Given that the Burgos basin in Mexico is an extension of the Texas Eagle Ford basin, there would appear to be 'geological promise' in that region (World Energy Council 2016: 23). Much will be dependent on Pemex's (the state-owed petroleum company) capacity to enter into contracts and form partnerships with foreign developers.

Europe

While some countries seek 'energy independence' from Russia (Dodge 2016: 97) through the domestic uptake of UGM, others, such as France, oppose UGM altogether and have, since 2011, imposed a total ban on fracking (Keeler 2016: 43). Belgium (since 2012), the Netherlands (since 2013 and extended for 5 years in 2015) and, Scotland and Wales (since 2015) have also introduced moratoria or bans, subject to further research (Vetter 2015).

In 2014, the European Commission established minimum principles for the exploration and production of hydrocarbons using fracking, effectively opening the door to UGM (European Commission 2014) and while the Aarhus Convention (on public participation and access to justice in environmental matters; Janjatović and Pokimica 2011) potentially impacts on UGM uptake in Europe European unconventional recoverable shale gas resources simply remain limited. They are estimated to be only 15 per cent of the world's total, with 4 per cent of that 15 per cent in Russia (Dodge 2016: 109).

UK

Cuadrilla has been at the forefront of UGM activity in England. However, a fracking moratorium between 2011 and 2012 slowed its progress. Originally the Lancashire County Council refused Quadrilla planning permission for UGM (Gray and McCoy 2016) but after a public inquiry the Secretary of State for Communities and Local Government overturned the decision. Anti-fracking campaigners sought judicial review of the Secretary's decision but in April 2017 the High Court dismissed the appeal.

Cuadrilla commenced exploratory drilling in August 2017 but local action groups have remained vocal in their opposition (DESMOGUK 2017).

In May 2016 Third Energy was granted the first planning approval for UG in England at its Yorkshire site (Delabarre et al. 2017).

Poland

Poland needs 'to drill 400 new shale gas wells per year for [it] to become self-sufficient(Dodge 2016: 113). Presumably state-ownership of UG would assist the pace of UGM uptake, by-passing the need for negotiations with individual land owners (ibid.: 115). However, given that Chevron, Exxon Mobil, Total and Marathon Oil have ceased UGM activities in gas rich Poland (ibid.: 128), UGM looks unlikely to provide the European energy 'solution'.

Germany

Opposition to German UGM remains high (German Advisory Council of the Environment 2013) particularly environmentally-based opposition in North-Rhine Westphalia and Lower Saxony where UG reserves are promising. Opposition is also strong from German beer brewers and arms of the Protestant Churches (Dodge 2016: 126).

Nevertheless, ExxonMobil commenced exploration activities and opened a dialogue involving independent experts from German research organizations. The dialogue concluded that despite new risks from fracking, UGM bans were unnecessary if caution were applied (Vetter 2015).

A framework that authorized only scientifically-supported testing measures and which operated as an effective ban on fracking was relaxed in 2014 and in 2015 the German Bundeskabinett introduced legislation permitting commercial fracking down to 3,000 metres depth. However, Germany's Energiewende (Energy Transition) is geared towards renewables rather than UGM and aims to 'produce 40–45% of its electricity from renewable sources' by 2025 (Dodge 2016: 126).

Australia

UGM growth in Queensland was stimulated by the Electricity Act 1994, which stipulated that Queensland electricity retailers had to source 13 per cent of their electricity from gas-fired generation, increasing to 15 per cent in 2011. In 2015–2016 alone, 685 CBM wells were drilled and 979 PJ were produced (DNRM 2017).

UGM activity further escalated with the decision to convert CBM to liquefied natural gas (LNG) for the export market – a world first. Three LNG plants are being developed on Curtis Island off Gladstone. Given that Queensland's remaining 2P reserves of CBM at 30 June 2016 were estimated to be 41, 229 PJ (DNRM 2017: 5) this project is likely to continue for some time.

In New South Wales there are estimated to be 3,082 PJ of 2P (UG) reserves but considerably more contingent resources (GeoSciences Australia in APPEA 2016: 7). The main NSW CBM projects include the Camden, Gloucester and Narrabri projects.

Meanwhile in Western Australia in 2017 (where there are 800 fracking wells), in Victoria[1] in 2017, and, in the Northern Territory in 2016, moratoria were introduced.

Shale gas mining is not yet well developed anywhere in Australia. However, if most of the known CBM reserves remain committed to the LNG export market, it is possible there will be increased pressure to exploit shale gas reserves.

Protest against and resistance to UGM remain robust in Australia despite the introduction of anti-protest legislation (Gray 2014; Ricketts 2017; Heath and Burdon 2017).

Unconventional gas and ecological integrity

The above update reflects one response to society's demands for more energy. It is simply to find new energy sources, such as UG, which will help feed a voracious, global energy appetite. This approach conceptualizes society's energy appetite within an economic growth paradigm. That paradigm supports expanding gross domestic products, further exploiting natural resources, and acquiescing to ever-increasing demands for more energy. Yet, seeing energy issues this way does nothing to supress the energy appetite and its deleterious effects. It simply defers the day when over-exploitation of natural resources has to be confronted. Further, continual economic growth contributes to global economic inequality (see Meadows et al. 1972) and does little to increase happiness (Victor 2008). There are, however, other ways of thinking about demands for energy.

Another way is through an ecological integrity frame. Ecological integrity values equally all components of the planet, including human beings, land, water and biota, for example. It also emphasizes the importance of planetary boundaries. Ecological integrity affirms 'an ecological ethic that values natural systems and the multiple forms of life' (Pimentel et al. 2000: 10). It also underpins concepts such as sustainability (Bosselmann 2014: 8).

If ecological integrity were used as a frame through which energy demands were mediated, it is likely that there would be a decrease in UGM activity. Having to incorporate non-human entities and beings into the community of justice presumably would de-emphasize some human priorities such as UG exploitation. There may also be a retreat from the market-based governance that presently dominates how we manage our relationship with each other and the planet. Instead of propertizing resources (such as UG) and creating the conditions for corporations and others to profit from UG exploitation, we may instead witness a shift towards alternative legal and management tools such as 'commoning'. An ecological integrity frame may also invite reliance on two companion concepts: (a) the common heritage of humankind (CHH) and (b) public trusteeship (PT).

The CHH includes the protection of our common natural heritage. As an international law principle CHH resists the creep of state sovereignty, has an intergenerational element and a wide ambit of protection. Operating in tandem with PT it may well serve to protect UG, land, water, biota and ultimately humanity.

In the context of UG, the concept of PT casts the state, as a trustee, holding UG (or land/water where UG is found) for the benefit of current and future generations. The trustee is under fiduciary and regular trust duties that require it does not alienate the resource into private ownership nor permit its injury or destruction. Indeed the state, as trustee, is under a duty to protect and steward the resource for the benefit of the public (see Sax 1970; Wood undated).

Application of PT in the UG space may necessitate many governments changing course from supporters to resisters of UGM.

In Pennsylvania, PT was applied in the fracking context. The legislature had passed an Act which sought to prevent a ban on, or regulation of fracking imposed by the local/municipal government. A plurality opinion of the Supreme Court held that the legislature had violated the public trust doctrine.[2] It also held that the Pennsylvania constitution did not create new public trust rights but rather enumerated rights that are 'inherent in man's nature'.[3]

It has since been argued that anti-fracking advocates could use the public trust doctrine to: (a) litigate enforcement of their government's public trust obligations, invoking the judiciary to protect their environmental rights; and/or (b) strengthen local anti-fracking ordinances by tying them to PT and 'expressing constitutional trust duties' (Wood undated). Both suggestions by-pass statutory environmental law to help achieve ecological integrity outcomes. PT may be a powerful tool to help 'disenfranchised citizens … the world over … enforce the right to a liveable ecosystem' (ibid.).

Such a use of PT harmonizes well with the concept of earth governance; a concept that valorizes the role of the citizen rather than the state, in developing processes for forming the collective will, by reliance on earth as the common reference point (Bosselmann 2015). An ecological integrity framework allows such approaches to come to the fore and guide decision-making.

Conclusion

The foregoing update provides an empirical overview that may be used as a backdrop or baseline picture for further research in fields including: mining, water and food law; climate change; economics; ecology and environmental management. It reveals that UGM (and fracking) activities are widespread but uneven. It also highlights a range of different conditions in different countries, outlines varying degrees of uptake and appetite for UGM and refers to different regulatory and governance approaches. Further, by detailing the ascendency of UGM in places such as the US, Australia, Canada, South Africa and China and simultaneously noting the types of potential harms and/or risks associated with UGM, the chapter underscores the potential *reach* of negative outcomes that may result from UGM. If potential harms are not well managed, the effects could be catastrophic.

Adoption of an ecological integrity framework for decision making and management of UG is explored and recommended. Such a framework would serve to highlight the consumerist and consumptionist trajectory that has been evident from at least the twentieth century onwards and which has helped increase energy demands and helped drive the push for new energy sources. With UG being embraced globally as a new energy source, it is likely that the environment will become further degraded, and planetary boundaries will be reached more rapidly than if renewable energy sources were relied upon. The CHH, in the form or our shared natural resources, is threatened by unsustainable choices and ways of living. An ecological integrity framework makes this realization stark and the public trust doctrine offers a way forward.

Notes

1 *Resources Amendment Legislation (Fracking Ban) Act* 2017.
2 *Robinson Township v. Pennsylvania*, 83 A.3d 901, 977 (Pa. 2013) (plurality opinion).
3 PA. CONST. art. I, § 27; *Robinson*, ibid, 83 A.3d at 1016 n.36

References

APPEA, 2016, 'Unconventional Gas in Australia', available at www.appea.com.au/wp-content/uploads/2017/02/Final-APPEA-Report-to-CoAG-Unconventional-Gas-in-Australia-2016.pdf.

Australian Senate, 2016, *Select Committee on Unconventional Gas Mining, Interim Report*, available at www.aph.gov.au/Parliamentary_Business/Committees/Senate/Gasmining/Gasmining/Interim%20Report/d01.

Bosselmann K., 2014, *Earth Governance*, Edward Elgar, Cheltenham.

Bosselmann K., 2015, 'The Rule of Law Grounded in the Earth', in L. Westra and M. Vilela (eds), *The Earth Charter, Ecological Integrity and Social Movements*, Routledge, Abingdon.

Brandt, A. R., Heath, G. A., Kort, E. A., O'Sullivan, F., Ptron, G, Jordaan, S. M., Tans, P., Wilcox, J., Gopstein, A. M., Arent, D., 2014, 'Methane Leaks from North American Natural Gas Systems', *Science* 343: 733–735.

Corrigan C. and Murtazashvili I., 2016, 'Fracking in Africa', in Y. Wang and W.E. Hefley (eds), *The Gobal Impact of Unconventional Gas Development*, 171–197.

Delabarre J., Ares E., Smith L., 2017, *Shale Gas and Fracking*, briefing paper no. 6073, House of Commons Library, London.

DESMOGUK, 2017, 'Drilling Begins at Cuadrilla's Lancashire Shale Gas Site', available at www.desmog.uk/2017/08/17/drilling-begins-cuadrilla-s-lancashire-shale-gas-site.

DNRM, 2017, 'Queensland's Petroleum and Coal Seam Gas', available at www.dnrm.qld.gov.au/__data/assets/pdf_file/0008/1237742/qld-petroleum-coal-seam-gas-2017.pdf.

Dodge R., 2016, 'Unconventional Drilling for Natural Gas in Europe', in Y. Wang and W. E. Hefley (eds), *The Global Impact of Unconventional Shale Gas Development: Economics, Policy and Interdependence*, Springer, Berlin.

Doyon J.A. and Bradshaw E., 2015, 'Unfettered Fracking: A Critical Examination of Hydraulic Fracturing in the United States', in Barak G. (ed.), in *The Routledge International Handbook of the Crimes of the Powerful*, Routledge, Abingdon.

EIA, 2016, 'Frequently Asked Questions', available at www.eia.gov/tools/faqs/faq.php?id=46&t=8.

European Commission, 2014, Recommendation of 22 January 2014, available at http://eur-lex.europa.eu/eli/reco/2014/70/oj.

German Advisory Council of the Environment, 2013, 'Fracking for Shale Gas Production', available at www.umweltrat.de/SharedDocs/Downloads/EN/04_Statements/2012_2016/2013_09_Statement_18_Fracking_for_Shale_Gas_Production.pdf%3F__blob%3DpublicationFile.

GISERA, 2014, 'Impacts of Coal Seam Gas Infrastructure Development on Agricultural Soil: A Case-Study in Southern Queensland, Australia', Annual International Meeting Conference paper, July, available at https://gisera.org.au/project/making-tracks-treading-carefully.

Grafton R.Q., Cronshaw I., Moore M. (eds), 2017, *Risks, Rewards and Regulation of Unconventional Gas: A Global Perspective*, CUP, Cambridge.

Gray J., 2014, 'Frack Off! Law, Policy, Social Resistance, Coal Seam Gas Mining and the Earth Charter', in L. Westra and M. Vilela (eds), *The Earth Charter, Ecological Integrity and Social Movements*, Routledge, Abingdon.

Gray J. and McCoy, D., 2016, 'The Common Good, Unconventional Gas Mining, and Public Health', in L Westra (ed.), *The Common Good and Environmental Governance for the Support of Life*, Cambridge Scholars, Newcastle upon Tyne.

Heath M. and Burdon P., 2017, 'Silencing of Activism in Australian Law', *Alternative Law Journal* 42(3): 190.

Herlihy, S, 2009, 'Trading Water for Gas: Application of the Public Interest Review to Coalbed Methane Produced Water Discharge in Wyoming', *Wyoming Law Review* 9: 455.

Janjatović T. and Pokimica N., 2011, 'Strategy for Implementing the Convention on Access to Information, Public Participation in Decision-making and Access to Justice in Environmental Matters – The Aarhus Convention', available at www.osce.org/serbia/89086?download=true.

Keeler, J, 2016, 'The Politics of Shale Gas and Anti-fracking Movements in France and the UK', in Y. Wang and W.E. Hefley (eds), *The Global Impact of Unconventional Shale Gas*, Springer, Berlin.

La Fleur D. and Sandiford S., 2016, *The Risk of Migratory Emissions Resulting from the Development of Queensland Coal Seam Gas*, University of Melbourne Energy Institute, available at www.tai.org.au/sites/defualt/files/Migratory_emissions_20170417-LowerRes.pdf.

Masters J.A., 1984, 'Deep Basin Gas Trap, Western Canada', *American Association of Petroleum Geologists' Bulletin* 63(2): 152–181.

Meadows D.H., Meadows D. L., Randers J. and Behrens W. III, 1972, *The Limits to Growth*, Universe Books, New York.

Meng Q., 2016, 'The Impacts of Fracking on the Environment: A Total Environmental Study Paradigm', in *Science of the Total Environment*, Oxford: Elsevier.

Natural Resources Canada, 2017, 'Alberta's Shale and Tight Resources', available at www.nrcan.gc.ca/energy/sources/shale-tight-resources/17679.

Pimentel D., Westra L. and Noss R. (eds), 2000, *Ecological Integrity*, Island Press, Washington, DC.

Ricketts, A, 2017, 'Anti-Protest Laws: Lock Up Your Nannas', *Alternative Law Journal* 42(2): 107.

Rockstrom J., 2009, 'Planetary Boundaries: Exploring the Safe Operating Space for Humanity', *Ecology and Society* 14(2): 32.

SAOGA, undated, 'Upstream Oil and Gas in South Africa', available at www.saoga.org.za/oil-gas-hubs/upstream-oil-gas-south-africa.

Sax, J, 1970, 'The Public Trust Doctrine in Natural Resources Law: Effective Judicial Intervention', *Mich L Review* 68: 471.

Scholes B., Lochner P., Schreiner G., Snyman-Van der Walt L., de Jager M., 2016, *Shale Gas Development in the Central Karoo: A Scientific Assessment of the Opportunities and Risks*, CSIR, Stellenbosch, available at http://seasgd.csir.co.za/wp-content/uploads/2016/12/SGD-Scientific-Assessment-Binder1_LOW-RES_INCL-ADDENDA_21Nov2016.pdf.

Underschultz J., 2017, 'Unconventional Gas' in Devasahayam S., Dowling K. and Mahapatra M. (eds), *Sustainability in the Mineral and Energy Sectors*, Taylor & Francis, Boca Raton, FL.

Vetter, A, 2015, 'Shale Gas in Germany – The Current Status', available at http://gfzpublic.gfz-potsdam.de/pubman/item/escidoc:1397928:5/component/escidoc:1398003/Vetter_ship_33.pdf.

Victor, P, 2008, *Managing Without Growth: Slower by Design, Not Disaster*, Edward Elgar, Cheltenham.

Wang H., Feng M., Ton X., Liu Z., Wu S., Li D., Wang B., Xie Y. and Yang L., 2016, 'Assessment of Global Unconventional Oil and Gas Resources', *Petroleum Exploration Development* 43(6): 925–940.

Willacy M., 2017, 'Coal Seam Gas Oversight Repeatedly Cut by Newman Government, Cabinet Documents Reveal' *ABC News* (online), 2 March, available at www.abc.net.au/news/2017-03-02/coal-seam-gas-oversight-repeatedly-cut-by-newman-government/8313632.

Wood M-C, undated, 'Enforcing Human Rights Against Fracking Through the Public Trust Principle', available at www.humansandnature.org/enforcing-human-rights-against-fracking-through-the-public-trust-principle.

World Energy Council, 2016, *World Energy Resources: Unconventional Gas, a Global Phenomenon*, World Energy Council, London.

Part III

New challenges to global governance

13 The reactionary turn in American environmental policy

The Trump effect

Sheila D. Collins

The election of Donald Trump in November 2016 to the presidency of the United States signalled not only the implosion of the liberal democratic state but a new and alarming threat to efforts to limit climate change and species extinction, not to mention the threat to the peace, health and safety of the world. The election of a man like Trump to such a critical office was totally unexpected by liberal political elites, but might have been foreseen if not for political blinders. Trump, however, is not the only worry for those who fear for the planet. The Republican majority in Congress is as much to blame. Let's just call it the Trump phenomenon.

In one respect, Trump's ascension can be seen as a logical outcome of forty or more years of neoliberal rule – a phenomenon similar to the rise of right-wing populist movements across Europe. Years of austerity, coupled with rapidly changing demographics, the rise of non-state terrorism and the failure of the Democratic Party to present a viable alternative vision and programme have led to an angry, bitter and xenophobic populist backlash by a white working class that was left behind by neoliberalism. The populist backlash is now firmly entrenched in the White House and in a segment of the congressional majority. How this happened can be explained by the nature of the United States's political system as well by as its particular cultural and political history.

The exceptional American political system

The United States is governed by a political system whose basic mechanisms were constitutionally enshrined in 1776 and which has rarely, and only after great difficulty, been amended ever since. Presidents are elected, not by popular vote, but by the Electoral College, an anachronistic mechanism that gives less populous states in both the Congress and Electoral College an advantage disproportionate to their populations. To further complicate the picture, electoral votes are counted through a winner-take-all system that in the last sixteen years has twice resulted in the presidency being won by a candidate who lost the popular vote.

Another peculiarity of the American political system is its weak party system. Despite the trappings of party organization, there is really very little of it.

A weak party system and divided government mean that there are multiple points of entry for outside interest groups who want to influence public policy through lobbying and campaign donations. One result is that money plays a much larger role in elections than it does in other countries. The Supreme Court has facilitated the role of big money in politics through three decisions which have declared that corporations are 'people' and endowed with free speech, and further, that political contributions constitute free speech. A third ruling has opened the floodgates of political contributions to well-heeled interest groups like the billionaire right-wing Koch brothers, who made their money in fossil fuels and have reportedly spent close to $20 million to elect Republican candidates who had to pledge to do nothing about global warming that would cost anything (PBS 2017).

A weak two- party system also results in the absence of political or ideological clarity when voters go to the polls. Since the two major parties have to represent the interests of the majority, their messages have to be so general that it is hard for the average voter to discern where their real interests lie. Moreover, the parties and candidates can say one thing in the primary races where their core constituencies are most active and change their message in the general election. It was this lack of clarity from the mainstream candidates and Trump's specificity and promise to carry out the policies that he ran on that helped to get him elected.

Another aspect of the American political system that has resulted in an entrenched reactionary congressional majority is the way in which congressional districts are drawn. Every ten years when the census is taken, the number of seats allotted to each state is reallocated according to population. Districts are then redrawn by the majority party in each state legislature. With deindustrialization, the population of the South has gained seats at the expense of the Northeast and industrial Midwest. This has given Southern politicians, who are predominantly Republican and conservative, an advantage in Congress and state legislatures. They have used their advantage to redraw political districts – a practice called Gerrymandering – to favour their own party giving them representation out of all proportion to the actual popular vote and have used their power to engage in voter suppression efforts aimed at people of colour, low income populations and the young, all of whom generally vote for the Democratic Party. Thus, the United States has both a president and a congressional majority who actually represent minority viewpoints. This is the main reason that the US ranked last among Western democracies in a 2017 index of voting fairness.[1]

Add to the above, a voting system with far too many elections that last far too long and make it as hard as possible for people to vote. Each state sets its own registration and voting rules. Thus, there are fifty different voting systems. There is no national holiday during which people can go to the polls, nor are elections held on weekends. With very little information to go by, and the difficulty of getting out to vote, it is no wonder that the US has one of the lowest voter turnouts in the industrialized world. This situation makes it easier for a determined, well-funded minority to grab an election.

Finally, in a non-parliamentary system like the United States, when the president and congressional majority represent the same party, the so-called constitutional checks and balances no longer work. With very few exceptions, the Republicans in Congress have gone along with Trump and seem willing to ignore the egregious violations of ethics and law that are pouring out of the White House for fear of losing their voting base. This has led to what is arguably the most corrupt administration in American history.

Underlying peculiarities of American political culture

Naomi Klein has commented that 'Trump is not a rupture at all, but rather the culmination – the logical end point – of a great many dangerous stories our culture has been telling for a very long time' (Klein 2017). American political culture has been deeply affected by slavery, structural racism and xenophobia. Unlike Germany, which confronted its fascist past, the United States has never confronted its racist history. As a result, throughout American history racism and xenophobia have reared their heads whenever economic conditions threaten the well-being of a white working class that has been conditioned to accept white supremacy as an entitlement of citizenship. Beginning with Richard Nixon's 'Southern Strategy', Republican politicians have used this fundamental cleavage to win votes through the use of racially coded language and images. Much of the opposition to Obama's presidency and policies was fuelled by racism, which has had the effect of aligning whites who have lost economic ground through neo-liberal globalization with the very class of people responsible for their economic decline.

Another peculiarity of American political culture is its excessive valorization of individualism. The old canard that one can 'pull themselves up by their own bootstraps' was always a myth but myths can have enduring political power. This mythology is most deeply embedded in rural areas of the country. The strength of this myth helps explain the weakness of American union power and the ability of politicians and business owners to so easily destroy the labour movement which was once the source of a more collective political consciousness.

A corollary to individualism is an aversion to the state, which has also been embedded in American political culture from the beginning. During the Great Depression of the 1930s Franklin D. Roosevelt ushered in the era of 'big government' and the Republican Party has exploited this fear of state power ever since, associating it with state coercion and a threat to individual liberty. More recently, powerful fossil fuel interests like the Koch brothers have exploited this fear, funding the Tea Party movement that has elected extreme Rightists to local, state and national office, fuelling the voter base for Trump. Trump played to this tune, portraying himself as a self-made businessman running against the Washington establishment who alone could solve the nation's problems.

Still another peculiarity of American political culture is its religiosity and the outsized role fundamentalist Christianity plays in American politics. Right-wing Christians played an important role in the election of Donald Trump.

The changing media environment

The US political system and culture alone could explain the rise of a populist leader, but they can't fully explain the particular type of populist leader represented by reality TV star Donald Trump without changes that have been occurring in the media environment over the last quarter century. Before cable news and the Internet, the US media was dominated by three giant media companies that presented a fairly standard national narrative representing the centre of the political spectrum. In general elections, Republicans ran slightly to the right of this centre and Democrats slightly to the left. All of that changed with the advent of the Internet, 24/7 cable TV, reality TV and social media and the rise of a powerful right-wing media empire. These changes have diminished the role of newspapers and news magazines to provide in-depth coverage of events and historical background. The US now has a bifurcated media framework. Liberals and progressives get their information and interpretation of events from one set of sources and conservatives and reactionaries from quite another. The result is a deeply polarized population.

Not only are there two very different sets of realities in which even the facts are disputed, but rumours, wild conspiracy theories, and outright fabrications now pass for 'news' in the right-wing media and the language of politics has been reduced and debased to the level of a 140-character tweet. History, context and critical analysis are found now only in a few liberal and progressive publications and cable TV shows. In the right-wing media, these are absent.

Even the liberal media, however, has a tendency to turn politics into theatre. Only the most sensational story will get coverage while important stories go ignored and evidence-based science gets a mixed record. The result is a significant sector of the electorate that has been robbed of its ability to think clearly and to reason effectively.

The Russian effect

Considerable evidence has been amassed that the Russian government manipulated the 2016 elections with the purpose of electing Trump, who had expressed admiration for Prime Minister Putin and whose campaign staff and family members had ties to Russian operatives. During the campaign, Russian computer operatives flooded social media with thousands of fake ads and comments from fake 'Americans', and repurposed videos meant to sow distrust of the federal government, of the electoral process and of once trusted institutions like the mainstream media, cleverly exploiting the country's existing political grievances, cultural resentments and news literacy (Collins, Poulsen and Ackerman 2017; Isaac and Shane 2016; Sanger and Shane 2016; Wakabayashi and Shane 2017). In the process the Democratic candidate was demonized and the former Democratic administration's accomplishments discredited. As of this writing, the investigation of the Russian effect is still in process, but absent the other factors that have been enumerated the Russians couldn't have done it by themselves. If they are implicated, the conditions for their influence had to be ripe.

Trump's assault on the environment: a preliminary assessment

While it is still too soon to tell what the net effect of Trump's policies on the environment will be he has already signalled that his will be the most environmentally destructive administration in modern American history. In addition to calling for drastic cuts to the budget of the Environmental Protection Agency, Trump's Office of Management and Budget called for cutting 25 percent of its staff. Calling climate change a hoax manufactured by the Chinese, Trump has appointed people to head the three agencies with most responsibility for overseeing the environment – the Environmental Protection Agency (EPA), the Energy Department and the Interior Department – who are all opponents of environmental regulations and sceptical if not deniers of climate change. Scott Pruitt, Trump's pick for EPA director had for years been leading the fight against Washington's environmental regulations. While serving as Oklahoma's attorney general, Pruitt had headed a group of over two dozen Republican attorneys general representing states heavily dependent on the fossil fuel industry. With funding from the Koch brothers, the body sued President Obama's EPA fourteen times in efforts to undo environmental regulations. Pruitt's mandate under Trump is to radically reduce the size and scope of the EPA and to reverse Barack Obama's climate change initiatives. On taking over the EPA Pruitt hired people with deep ties to the fossil fuel industry, in effect making them the regulators of the companies they served.

In addition to hiring people hostile to the mission of the agency they serve, the Trump administration has given speeches and publicity that ignore or contest the agency's basic mission and that pledge overt allegiance to regulated industries. Career employees have been marginalized, monitored and silenced. The result is that morale has plummeted and many express deep anxiety about their own careers and the future of environmental protection and the EPA (Sellers et al. 2017). As Eric Shaffer, Director of the EPA's office of Civil Enforcement under the Obama administration put it: 'What it [the appointment of Pruitt] conveyed is that this is a hostile takeover. You, the scientists and lawyers and engineers at the agency are no longer valued. This is a political operation' (PBS 2017).

Rick Perry, head of the Department of Energy, is a former governor of Texas who at one time called for the dismantling of the Energy Department. He supports remaining in the Paris Climate Agreement but would like to 'renegotiate' it. His actions, however, have come down on the side of regulatory relief for the fossil fuel industry. Ryan Zinke, who heads Trump's Department of the Interior once believed in climate change and supported renewable energy, but since joining the Trump administration has changed his tune. He has called the Paris Climate Agreement a bad deal, unfair to the US, and, at any rate one that would not make much difference for the climate and has taken steps to eliminate protections for public lands and wildlife habitat, opening parts of national monuments protected by the Obama administration to mining interests (Geiling 2017).

The Trump administration's assault on the environment falls into three broad categories: first are the threats to science and science-based research; second the defunding of domestic environmental agencies and programmes; and third is the withdrawal of the commitment to global environmental governance.

Threats to science-based research

On taking office, Trump issued an Executive Order requiring each agency in the federal government to enact a 'regulatory reform' agenda, whose real purpose is to undo as many regulations on business as possible. Since then there have been multiple executive orders requiring the EPA not just to favour fossil fuels but to rescind two existing rules for every new one (with assessments based only on compliance costs and not on calculated benefits); to re-evaluate the rest of agency rules for 'burdensomeness'; and to reorganize with a view to downsizing (Sellers et al. 2017). According to a survey conducted by the *New York Times*, in October 2017, the Trump administration has sought to reverse more than 50 environmental rules (Popovich and Albeck-Ripa 2017).

Several government agencies have removed scientists from their agencies and scientific data from their websites. For example, the EPA removed the main climate page from its website and the word, 'science' from the mission page of its Office of Science and Technology. It has also removed over 200 climate webpages for state, local and tribal governments making it harder for them to access information critical to helping them adapt to and mitigate the harms of climate change (Rinberg et al. 2018). Similarly, Rick Perry ordered staff of the Energy Department's Office of International Climate and Clean Energy not to use the phrases 'climate change', 'emissions reduction' or 'Paris Agreement' in their reports. In addition, Pruitt issued an order that half the expert members of a board that advises the Environmental Protection Agency on the integrity of its science were not to be reappointed (Biesecker 2017).

Trump can get away with all this because he has a compliant congressional majority. In its early decades, the E.P.A., with the exception of the Reagan administration, enjoyed bipartisan congressional support. This is no longer the case. Emboldened by the takeover of Congress by a Republican majority in 1994, the fossil fuel industry saw a way to undo the environmental regulations they had chafed under since the establishment of the E.P.A. in 1970.

Threats to public health and safety

A climate change denying Republican congressional majority that now has the upper hand in Washington introduced a bill misleadingly called the "Honest and Open New E.P.A. Science Treatment Act of 2017 (H.R. 1430)." If enacted, the bill would obstruct the E.P.A.'s use of important scientific studies in essential agency work, forcing the E.P.A. to make determinations without access to entire categories of research that are essential to sound decision-making. The restrictions go against widely accepted guidelines for scientific research into environmental exposures and their health effects and will result in risks to the health and safety of citizens (Underhill et al. 2017). The bill was passed by the House but has yet to be passed by the Senate.

Coming into office declaring that he was going to rescue coal, Trump declared President Obama's Clean Power Plan, his signature policy to curb greenhouse gas emissions from power plants, dead in the water; and in October 2017 Scott Pruitt formally repealed the rule which was expected to cut power sector emissions 32 percent by 2030, relative to 2005. Eliminating the Clean Power Plan makes it less likely that the United States can fulfil its promise as part of the Paris climate agreement as well as adding to the health risks of the public. Dozens of other actions taken by the Trump administration put the health and safety of the public in grave new jeopardy, among them: reversing a ban on agricultural use of the toxic pesticide chlorpyrifos and curbing requirements for providing public information on workplace risks, especially with regard to chemical processing and information access for first responders to industrial accidents.

These measures do not even begin to match the magnitude of Trump's new threats to global health and the environment from his decision to increase the military budget, upgrade the nation's nuclear arsenal and use bellicose language to threaten other nuclear states like North Korea.

Funding cuts for domestic environmental protection and climate change mitigation

Mick Mulvaney, the director of Trump's Office of Management and Budget, said at a White House briefing in March 2017 that 'As to climate change … We consider that to be a waste of your money to go out and do that' (Fountain and Schwartz 2017). Trump's proposed 2018 budget, slashes the Environmental Protection Agency's budget by 31 percent – a steeper cut than any other agency. Environmental science would be especially hard hit. The Office of Research and Development would lose 48 percent of its funding, the Science Advisory Board budget would be cut by 84 percent; with severely reduced capabilities to review EPA research programmes and provide scientific advice to the agency, and the Superfund budget for toxic waste clean-up would be cut by 30 percent. Fifty-six programmes are slated to be cut entirely, including those for climate change, pesticide safety, environmental justice, environmental education, indoor radon, radiation, and many regional programmes such as the Great Lakes programme (Rinberg et al. 2018).

Assault on Global Environmental Governance

In keeping with his campaign slogan of 'America First', Trump has taken several steps to withdraw the United States from its historic commitment and contributions to global governance. The Departments of State, Energy, and the EPA have removed information about the federal government's international obligations regarding climate change, downplaying U.S. involvement (Rinberg et al. 2018). Beyond Trump's announcing his intention to withdraw from the Paris Climate Agreement, 'nearly every possible funding item that could support international climate action has been identified and cut', according to the World Resources Institute. 'But the administration has gone further, with accompanying language in the request making it clear they have no intention of contributing climate finance in future years' (Thwaites 2017). However flawed that past commitment has been, the US retreat poses new threats to the global environment, global health and peace. According to one U.N. official, Trump's huge proposed cuts in US funding for the U.N. would make it 'simply impossible' for it to carry out essential work to promote peace, combat poverty and provide humanitarian assistance (Lederer 2017).

The Trump effect: a preliminary prognosis

Despite the apparent juggernaut of the Trump onslaught there has been considerable pushback to his policies on the part of scientists, environmentalists, peace and human rights activists and concerned citizens. During Trump's first 100 days several massive demonstrations took place in Washington and around the country by citizens expressing their outrage at his proposed policies and a new political movement aimed at running anti-Trump candidates for office was born. While there is little hope that the current Congress will restore many of the cuts Trump has ordered, there will certainly be legal challenges and procedural delays to the implementation of his policies. Trump can't immediately withdraw from the global Climate Agreement and experts say it may take him to the end of his term to fully disengage. Environmental groups and several states plan to challenge the repeal of several of Trump's rule reversals in federal courts.

Alarmed by the anti-science ideology of the Trump administration and Republican Congress, a small group of concerned researchers in the weeks after the US elections formed the Environmental Data and Governance Initiative (EDGI) whose mission is to preserve publicly accessible and potentially vulnerable scientific data and archive web pages from government agencies. It has since grown into an international network of over 100 people from thirty institutions. EDGI projects include monitoring and analysing changes to federal environmental agency websites, and researching the effects of environmental deregulation and changes at federal agencies.

Trump's agenda may yet be slowed by a growing movement among subnational actors. Former New York City Mayor Michael Bloomberg and California Governor Gerry Brown have been aggregating the emissions reductions submitted by the mayors of 227 cities and counties, nine states and about 1,650 businesses, universities and investors. Together, they submitted a plan to the United Nations pledging to meet the United States' greenhouse gas emissions targets under the Paris Climate Accord, despite Trump's decision to withdraw. In addition, Mr. Bloomberg's charitable organization has offered to donate $15 million over the next two years to help fund the budget for the International Framework Convention on Climate Change should it be needed – a figure representing the United States's share.

There is one last source of hope for the planet and that is in the trajectory of market forces and technological advances which are driving down the cost of solar and wind making them more competitive with fossil fuels. According to the IEA, nearly 60 percent of all new power generation capacity to 2040 will come from renewables and, by 2040, the majority of renewables-based generation will be competitive without any subsidies. Almost half the coal plants in the US have announced in recent years that they will close or switch to cleaner fuels. The market is simply no longer there. On a global scale, more than half the investment in new electricity generation is going into renewables. Even heavily Republican states like Texas and Iowa now lead the nation in wind power. Despite the absence of the Clean Power Plan, many utilities across the country have opted to shift to renewables, driven by cost concerns and state-level policies. States like California and New York, are already moving ahead of the targets set by the Clean Power Plan (Friedman and Plumer 2017).

Note

1 The index was created by the Electoral Integrity Project, an academic collaboration between the University of Sydney, Australia, and Harvard University's John F. Kennedy School of Government in Cambridge, Massachusetts.

References

Biesecker, Michael (2017) 'Trump Administration Hollows Out EPA Science Integrity Board', 8 May, www.apnewsarchive.com/2017/Trump_administration_hollows_out_EPA_science_integrity_board/id-88e8afe3170d49dd85ea8d84df708ecb.

Collins, Ben, Kevin Poulsen, and Spencer Ackerman (2017) 'Exclusive: Russians Impersonated Real American Muslims to Stir Chaos on Facebook and Instagram', *Daily Beast* (27 September), www.thedailybeast.com/exclusive-russians-impersonated-real-american-muslims-to-stir-chaos-on-facebook-and-instagram.

Fountain, Henry, and John Schwartz (2017) 'Scientists Bristle at Trump Budget's Cuts to Research, *NY Times* (16 March), www.nytimes.com/2017/03/16/climate/trump-budget-science-research.html.

Friedman, Lisa, and Brad Plumer (2017) 'EPA Announces Repeal of Major Obama-Era Carbon Emissions Rule', *New York Times* (9 October), www.nytimes.com/2017/10/09/climate/clean-power-plan.html?_r=0.

Geiling, Natasha (2017) 'Zinke Defends Trump's Climate Cuts by Getting Facts Wrong', *Think Progress* (8 June), https://thinkprogress.org/zinke-budget-hearing-climate-change-falsehoods-7b0253272ddd.

Isaac, Mike, and Scott Shane (2016) 'Facebook to Deliver 3,000 Russian-Linked Ads to Congress on Monday', *New York Times* (1 October).

Klein, Naomi (2017) 'Daring to Dream in the Age of Trump', *The Nation* (13 June), www.thenation.com/article/daring-to-dream-in-the-age-of-trump/?utm_source=Sailthru&utm_medium=email&utm_campaign=weekly%2006162017&utm_term=weekly.

Lederer, Edith M. (2017) 'Trump's Proposed Funding Cuts Would Make the UN's Work "Impossible"', *Time Magazine* (24 May), http://time.com/4793602/trump-un-funding-cuts-budget.

PBS (2017) *War on the EPA*, Frontline documentary, aired by PBS on 11 October https://www.pbs.org/wgbh/frontline/film/war-on-the-epa/.

Popovich, Nadia, and Livia Albeck-Ripa (2017) '52 Rules on Way Out Under Trump', *New York Times* (6 October), www.nytimes.com/interactive/2017/10/05/climate/trump-environment-rules-reversed.html.

Rinberg, Toly, Maya Anjur-Dietrich, Marcy Beck, et. al. (2018) "The First 100 Days and Counting, Part 3: Changing the Digital Climate," Environmental Data and Governance Initiative, http://100days.envirodatagov.org/changing-digital-climate/#h.2xw111cdg70r.

Sanger, David E., and Scott Shane (2016) 'Russian Hackers Acted to Aid Trump in Election, US Says', *New York Times* (9 December), www.nytimes.com/2016/12/09/us/obama-russia-election-hack.html?rref=collection%2Fnewseventcollection%2Frussian-election-hacking&action=click&contentCollection=politics&region=rank&module=package&version=highlights&contentPlacement=3&pgtype=collection.

Sellers, Christopher, (2017) 'The First 100 Days and Counting, Part 1: The EPA Under Siege, Executive Summary' Environmental Data and Governance Initiative, https://100days.envirodatagov.org/epa-under-siege/. https://www.pbs.org/wgbh/frontline/film/war-on-the-epa/.

Thwaites, Joe (2017) 'Trump Budget Nickel-and-Dimes America – and the World', World Resources Institute, May 23, 2017, www.wri.org/blog/2017/05/trump-budget-nickel-and-dimes-america-and-world.

Underhill, Vivian, Megan Martenyi, Sarah Lamdan, and Andrew Bergman (2017) 'Public Protections under Threat at the E.P.A.', Environmental Data and Governance Initiative, March, https://envirodatagov.org/wp-content/uploads/2017/03/Public-Protections-under-Threat-at-the-EPA.pdf.

Wakabayashi, Daisuke, and Scott Shane (2017) 'Twitter, With Accounts Linked to Russia, to Face Congress Over Role in Election', *New York Times* (27 September), www.nytimes.com/2017/09/27/technology/twitter-russia-election.html.

14 Moving from environmental law to ecological law

Frameworks, priorities and strategies

Geoffrey Garver

Introduction

The ecological economics community for several decades has been developing a counter-narrative to the conventional, growth-insistent economics underlying the current dominant orientation of human societies at all scales, local to global. The core vision of ecological economics is an economy that allocates and distributes goods and services strictly within scientifically-derived ecological limits (Daly 1996). This vision implicitly encompasses a need for change in the legal and governance systems that support the dominant economic regime. Planetary boundaries (Rockström et al. 2009; Steffen et al. 2015) are a compelling representation of global ecological limits that allow 'the flexibility to choose a myriad of pathways for human well-being and development' (Rockström et al 2009). This hopeful framing of the implications of an ecologically limited future invites a rigorous analysis of the need to shift from environmental law, which aligns with conventional economics, to ecological law, which aligns with the vision of a community of life – including humans – thriving within ecological limits.

The planetary boundaries framework is a response to the growing risk that the collective impacts of human activity will, unless significantly changed, tip the Earth's ecosystems into a state that will drastically limit their capacity to support human and other life (Rockström et al. 2009). Planetary boundaries research points increasingly to an urgent need to develop 'novel and adaptive' governance built on global ecological limits (ibid.). This chapter elaborates how current environmental law impedes pathways toward law and governance systems that would contain human activities within global ecological limits, lays out some basic elements of ecological law (still mostly conceptual) built around those limits and summarizes some challenges and opportunities in the hoped-for transition from environmental to ecological law.

The problem: what is wrong with environmental law?

Environmental economics attempts to correct environmental market failures by internalizing environmental costs into the prices of goods and services, for example by applying the 'polluter pays' principle (Brown and Garver 2009).

However, it operates within a neo-classical, growth-insistent economic paradigm that has so far been unable to integrate, let alone respect, critical aggregate ecological boundaries on human endeavours. Environmental law has yielded many important improvements in environmental quality since its modern inception in the 1960s and 1970s. Yet, it remains reductionist and fragmented, in that it is still largely focused on the individual sources of pollution without adequate measures to address aggregate impacts (Garver 2013). Provisions of environmental law that require consideration of broad, integrated implications of human activities, such as environmental impact assessment law, are typically limited (e.g. when they apply only to actions implicating governmental involvement or approval) and procedural (in that they do not require adoption of the most environmentally favourable alternative; ibid.).

As well, environmental law relies too much on flawed methods to monetize environmental harms and compare those monetized costs against other monetized measures of social welfare, all from a perspective that is too anthropocentric. The tendency to incorporate the language of money into the law is rooted ultimately in the conception of humans as apart from nature, and as rational actors free to accommodate and own elements of nature in the quest to maximize personal wealth and well-being (Nadeau 2006). Finally, environmental law overly expresses confidence that technological solutions will eventually emerge to solve whatever environmental challenges, such as climate change, humanity will confront.

The European Union's unsuccessful effort to implement the Convention on Long-range Transboundary Air Pollution (LRTAP) within ecologically derived limits is illustrative. The LRTAP adopts an ecologically-grounded approach to regulate air pollutant emissions based on scientifically based limits on the capacity of the ecosystems that receive those emissions to absorb them without risk of ecological disruption. In this approach, critical levels (emissions from sources) are 'concentrations of pollutants in the atmosphere above which direct adverse effects on receptors, such as human beings, plants, ecosystems or materials, may occur according to present knowledge' (UNECE undated). Critical loads (capacity of receptor ecosystems) are a 'quantitative estimate of an exposure to one or more pollutants below which significant harmful effects on specified sensitive elements of the environment do not occur according to present knowledge' (ibid.).

The critical loads and levels approach uses complex modelling to estimate the emissions reductions and allocations that are needed to keep levels of individual pollutants or pollutants in combination below systemic thresholds (Umweltbundesamt 2004). The EU has applied the critical loads/critical levels approach to key air pollutants, including sulphur and nitrogen-based pollutants and heavy metals such as cadmium, lead and mercury, as well as to the combined effects of at least some pollutants (ibid.). Although dynamic modelling that takes into account both short and long term effects of air pollutants on receptor ecosystems has reasonably described source-receptor relationships with respect to pollutants like sulphur dioxide that cause acidification, it has been much less successful with respect to the more complex biogeochemical processes and feedbacks involved in eutrophication resulting primarily from nitrogen-based air pollutants (Spranger et al. 2008).

Beyond the challenges of establishing ecological standards for a wide range of pollutants in receiving ecosystems and in modelling source-receptor relationships, the EU's critical loads/critical levels approach has not prevented exceedances of ecologically derived limits based on this approach (Spranger et al. 2008; Pelletier 2010). Indeed, the EU has consistently found that preventing such exceedances is not anticipated in any concrete time frame (European Union 2001; Umweltbundesamt 2004; Spranger et al. 2008; Gothenburg Protocol 2011). Technical and economic factors continue to be predominant and to impede progress toward those limits, as they have with unsuccessful efforts to achieve unenforceable, elusive goals of environmental laws in the United States and elsewhere (Pelletier 2010).

The compelling narrative that emerges from the planetary boundaries concept, which is based on long-term maintenance of safe operating space for humanity, has had virtually no impact on environmental law at any level. The Paris Agreement, built around voluntary commitments of the signatory nations, would not result in reductions of greenhouse gas emissions that, according to the overwhelming weight of scientific evidence and analysis, would be needed to avoid disruption of the Earth's climate that would be catastrophic (Rogelj et al. 2016). A rigorous framework for law and governance that would halt and eventually reverse humanity's devastation of biodiversity at all scales, local to global, remains elusive. Sweden has set an example by trying to develop details as to how the planetary boundaries approach could be implemented, but that work has not yet produced hard results (Nyquist 2013). The 'green growth' concept (UNEP 2011), grounded in the belief that perpetual increases in the efficiency in the use of material and energy resources, along with technological innovations, will sufficiently counter ongoing ecological impacts of perpetual growth with regard to challenges such as climate change, biodiversity loss and nutrient pollution, lacks scientific rigour.

The solution: what is ecological law?

Table 14.1 summarizes key distinctions between growth-insistent environmental law and limits-insistent ecological law. Ecological law is grounded in the concept that humans are integral members of Earth's community of life who are unique in having created a global ecological crisis that calls for rapid transition to new norms, behaviours and ultimately enforceable rules that keep the collective human enterprise within critical ecological limits. Like ecological economics, ecological law is still mostly conceptual and not yet widely understood or practised, largely because they both envision a transition away from the hard-wired insistence on economic growth that undergirds policy and decision making globally. Yet, the growth-insistent economic model and the legal and governance systems that support it are entirely socially constructed, and therefore subject to

Table 14.1 Distinguishing ecological law from environmental law.

Feature	*Environmental law*	*Ecological law*
Human–nature relationship	Humans are separate from and superior to nature; goal is perpetual progress in human control of nature, with strong reliance on technological solutions (e.g. geo-engineering to mitigate climate change)	Humans are a part of nature; goal is a mutually enhancing human–Earth relationship, with humility as to prospects for technology to solve complex ecological challenges
Enforceable environmental or ecological limits	Enforceable limits on pollution and development are mostly reductionist, end-of-pipe, subordinate to economic growth, and subordinate to property rights	Ecological limits have primacy over social and economic spheres, and are based on a holistic, integrated, systems-based understanding of the human–Earth relationship; open to de-growth/steady state economics
Use of materials and energy	Promotes efficiency, with a core faith in perpetual decoupling of energy and material throughput and consequent impacts from perpetual economic growth	Promotes sufficiency and drastic reduction in material and energy throughput to keep economy within ecological bounds
Scale	Strong commitment to state sovereignty: weak international/global regimes to address ecological challenges; global and regional trade rules encourage competition and impede strong domestic regimes for environmental protection	Core commitment to subsidiarity principle: global regime with enforceable supranational rules for global ecological issues, with preference for local regulation and respect for local regimes per Ostrom's criteria (Ostrom 1990)
Fairness	Core belief in fairness of markets, with some need for correction (e.g. polluter pays principle, internalization of environmental externalities, etc.); tendency to monetize values in decision making, e.g. monetary valuation of 'ecosystem services'	Strong limits on market mechanisms as needed to respect ecological limits; bigger role for non-market decision-making; focus on ensuring interhuman, interspecies and intergenerational fairness; multi-criteria valuation methods preferred for decision making involving incommensurate values
Research, monitoring and adaptation	Effects on human health paramount; environmental effects studied but not determinative; weak precautionary approach; few mechanisms to adjust rules based on monitoring	Planetary boundaries and 'safe operating space' are key basis of research, monitoring and adaptation; strong precautionary approach

change as the flaws in their conceptual foundations become more and more clear. Ecological economics and ecological law are emerging social constructions that respond to the increasingly apparent impossibility of perpetual economic growth on our finite planet (even if economic growth is more and more decoupled from throughput of material and energy in the economy and consequent ecological impacts; Garver 2013).

Thomas Berry's conception of a mutually enhancing human–Earth relationship (Berry 1999) provides a compelling core objective for ecological economics and ecological law. Berry described a mutually enhancing human–Earth relationship as one that reflects that 'in reality there is a single integral community of the Earth that includes all its component members whether human or other than human [each of which] has its own role to fulfil, its own dignity, its inner spontaneity' (ibid.: 4). It is a relationship in which '[e]very being enters into communion with other beings' (ibid.). Berry aligns the transition to a mutually enhancing human–Earth relationship with the emergence of the Ecozoic Era, 'the period when humans will be present to the planet as participating members of the comprehensive Earth community' (ibid.: 8). With regard to law, Berry wrote that '[e]cology is not part of the law; law is an extension of ecology' (ibid.: 84) – in other words, that law should reflect and maintain a human role within the broader community of life that is life-enhancing and respectful of the ecological role of other members of that community. He argued that '[t]o achieve a viable human–Earth situation a new jurisprudence must envisage its primary task as that of articulating the conditions for the integral functioning of the Earth process, with special reference to a mutually enhancing human–Earth relationship' (ibid.: 61).

Merchant (2003) frames the notion of a mutually enhancing human–Earth relationship in a way that avoids the flaws of a pure trust in both (1) human control and domination of nature and (2) a vision of ideal nature that ultimately excludes humanity. She frames this middle path in terms of a partnership between humans and Earth, asserting '[a] partnership ethic holds that the greatest good for the human and nonhuman communities is in their mutual living interdependence' (ibid.: 223). This partnership ethic rejects the hubris of the Enlightenment project of human mastery over nature (ibid.). It extends moral standing, equity and rights of inclusion and participation to nonhuman communities and species, as well as to all human communities and cultures (ibid.). Although Merchant's partnership ethic leaves open questions as to the scope of notions such as equity and health, it broadly aligns with Berry's notion of a mutually enhancing human–Earth relationship. The conceptual core of hers and Berry's ethical vision necessarily implies that a notion such as ecosystem services (to humans) be complemented with the notion of human services to ecosystems. This notion of mutual service has virtually no firm toehold in contemporary law and policy. Yet, it must be part of a practicable approach to ecological law.

Human-inclusive ecological integrity is another powerful concept at the heart of ecological law. Many international agreements or soft law instruments refer to ecological integrity as an overarching, or at least significant, objective (Kim and Bosselmann 2013). For example, a core principle in the Earth Charter is to 'protect and restore the integrity of Earth's ecological systems, with special concern for biological diversity and the natural processes that sustain life' (Earth Charter 2016).

To serve as the foundation for an overarching, global objective that can be applied at the scale of landscapes or the entire Earth, ecological integrity must encompass not only the most pristine wild ecosystems, but also dense human settlements and other areas in which humans or their impacts have significantly transformed the evolutionary trajectory of the pre-human or an imagined human-free ecosystem (Ellis and Ramankutty 2008). A rigorous yet practicable notion of ecological integrity must incorporate some level of symbiosis between humans and non-human nature in order to be consistent with a mutually enhancing human–Earth relationship. The challenge becomes determining where, along the spectrum from the least impacted to the most anthropogenically transformed ecosystems, benchmarks for ecological integrity that are practicable in law and governance systems can and should be drawn at different spatial and temporal scales, accounting for humans as an integral ecosystem component. This determination necessarily reflects the aggregate of the individual and collective intentions of human beings; human intention is thus a critical variable for the human prospect. Ecological law is ultimately a hopeful vision of human societies orientating this intention so as to develop workable norms and rules for a thriving human community within a thriving community of all life on Earth.

Transition: how do we move from environmental to ecological law?

Significant obstacles stand in the way of a transition to a limits-insistent global legal system that promotes a mutually enhancing human–Earth relationship. The theoretical notion of law that this chapter presents is a future vision not only of law, but also of the social, cultural, political and economic contexts in which law is embedded (M'Gonigle 2008). It is a vision that implies an inevitable evolution away from current political orders and power structures, toward new ones that must emerge as this transition unfolds. To attain this vision, the resistance to change in the legal system and related systems is high and will take a long time to overcome. This is particularly true with respect to necessary paradigm shifts – especially the transition from the dominant growth-insistent paradigm to the limits-insistent paradigm proposed in this chapter. This transition implies a significant change in deeply entrenched power structures and political orders at all scales – change that will likely emerge in unpredictable ways and through unpredictable actors (Brown and Garver 2009).

The creation of the Ecological Law and Governance Association (ELGA) in 2016, based on the Oslo Manifesto (2016), was an important step toward the transition from environmental to ecological law. The Oslo Manifesto states:

> To overcome the flaws of environmental law, mere reform is not enough. We do not need more laws, but different laws from which no area of the legal

> system is exempted. The ecological approach to law is based on ecocentrism, holism, and intra- /intergenerational and interspecies justice. From this perspective, or worldview, the law will recognise ecological interdependencies and no longer favour humans over nature and individual rights over collective responsibilities. Essentially, ecological law internalizes the natural living conditions of human existence and makes them the basis of all law, including constitutions, human rights, property rights, corporate rights and state sovereignty.
>
> (Oslo Manifesto 2016: para. 5)

ELGA is a growing network of jurists and others who are convinced of the need to further develop ecological law and to seek opportunities to put it into practice.

Conclusion

The transition from environmental to ecological law is a long-term project. It implicates a radical transformation in humanity's common understanding of the human–Earth relationship and of realistic pathways toward making that relationship mutually enhancing. Hopefully, this chapter lays out the broad principles underlying this much needed transition.

References

Berry, T. 1999. *The Great Work: Our Way Into the Future*. New York: Three Rivers Press.

Brown, P. G., and G. Garver. 2009. *Right Relationship: Building a Whole Earth Economy*. San Francisco, CA: Berrett-Koehler.

Daly, H. E. 1996. *Beyond Growth*. Boston, MA: Beacon Press.

Earth Charter. 2016. Earth Charter Initiative. Available at http://earthcharter.org.

Ellis, E. C. and N. Ramankutty. 2008. Putting People in the Map: Anthropogenic Biomes of the World. *Front Ecol Environ* 6(8): 439–447.

European Union. 2001. Directive of the European Parliament and of the Council 2001/81/EC of 23 October 2001 on national emission ceilings for certain atmospheric pollutants, O.J. L309/22.

Garver, G. 2013. The Rule of Ecological Law: The Legal Complement to Degrowth Economics. *Sustainability* 5(1): 316–337.

Gothenburg Protocol. 2011. Gothenburg Protocol to the 1979 Convention on Long-range Transboundary Air Pollution to Abate Acidification, Eutrophication and Ground-level Ozone (30 November 1999). Available at www.unece.org/fileadmin/DAM/env/lrtap/full%20text/1999%20Multi.E.Amended.2005.pdf.

Kim, R. E., and K. Bosselmann. 2013. International Environmental Law in the Anthropocene: Towards a Purposive System of Multilateral Environmental Agreements. *Transnational Environmental Law* 2: 285–309.

Merchant, C. 2003. *Reinventing Eden: The Fate of Nature in Western Culture*. New York: Routledge.

M'Gonigle, M. 2008. Green Legal Theory: A New Approach to the Concept of Environmental Law. *Neue Konzepte* 4: 34–38.

Nadeau, R. L. 2006. *The Environmental Endgame: Mainstream Economics, Ecological Disaster, and Human Survival*. New Brunswick, NJ: Rutgers University Press.

Nyquist, B., A. Persson, F. Moberg, L. Persson, S. Cornell and J. Rockström. 2013. *National Environmental Performance on Planetary Boundaries: A Study for the Swedish Environmental Protection Agency*. Stockholm: Swedish Environmental Protection Agency.

Oslo Manifesto. 2016. Oslo Manifesto. Adopted at the IUCN WCEL Ethics Specialist Group Workshop, IUCN Academy of Environmental Law Colloquium, University of Oslo, 21 June. Available at www.harmonywithnatureun.org/content/documents/20170404OSLO_MANIFESTO.pdf.

Ostrom, E. 1990. *Governing the Commons: The Evolution of Institutions for Collective Action*. New York: Cambridge University Press.

Pelletier, N. 2010. Of Laws and Limits: An Ecological Economic Perspective on Redressing the Failure of Contemporary Global Environmental Governance. *Global Environmental Change* 20: 220–228.

Rockström, J., W. Steffen, K. Noone, Å. Persson, F. S. Chapin, III, E. Lambin, T. M. Lenton, M. Scheffer, C. Folke, H. Schellnhuber, B. Nykvist, C. A. De Wit, T. Hughes, S. van der Leeuw, H. Rodhe, S. Sörlin, P. K. Snyder, R. Costanza, U. Svedin, M. Falkenmark, L. Karlberg, R. W. Corell, V. J. Fabry, J. Hansen, B. Walker, D. Liverman, K. Richardson, P. Crutzen, and J. Foley. 2009. Planetary Boundaries: Exploring the Safe Operating Space for Humanity. *Ecology and Society* 14(2): 32.

Rogelj, J., M. den Elzen, N. Höhne, T. Fransen, H. Fekete, H. Winkler, R. Schaeffer, F. Sha, K. Riahi, M. Meinshausen. 2016. Paris Agreement Climate Proposals Need a Boost to Keep Warming Well Below 2°C. *Nature* 534: 631–639.

Spranger, T., J.-P. Hettelingh, J. Slootweg, M. Posch. 2008. Modelling and Mapping Long-term Risks Due to Reactive Nitrogen Effects: An Overview of LRTAP Convention Activities. *Environmental Pollution* 154(3): 482–487.

Steffen, W., K. Richardson, J. Rockström, S. E. Cornell, I. Fetzer, E. M. Bennett, R. Biggs, S. R. Carpenter, W. de Vries, C. A. de Wit, C. Folke, D. Gerten, J. Heinke, G. M. Mace, L. M. Persson, V. Ramanathan, B. Reyers, S. Sörlin. 2015. Planetary Boundaries: Guiding Human Development on a Changing Planet, *Science* 347(6223): 1259885-1–1259885-10.

Umweltbundesamt. 2004. *Manual on Methodologies and Criteria for Modelling and Mapping Critical Loads and Levels and Air Pollution Effects, Risks and Trends*. Berlin: Umweltbundesamt.

UNECE (undated). ICP Modelling and Mapping: International Cooperative Programme on Modelling and Mapping of Critical Loads and Levels and Air Pollution Effects, Risks and Trends. Available at www.unece.org/env/lrtap/WorkingGroups/wge/mapping.htm.

United Nations Environment Programme (UNEP). 2011. *Towards a Green Economy: Pathways to Sustainable Development and Poverty Eradication – A Synthesis for Policymakers*. Nairobi: UNEP. Available at www.unep.org/greeneconomy/GreenEconomyReport/tabid/29846/language/en-US/Default.aspx (accessed 2 November 2014).

15 Achieving traction for ethics in environmental policy-making

Donald A. Brown

Introduction

This chapter will first review recent evidence that despite continuous growth in academic interest in environmental ethics since the 1970s, and not withstanding that global environmental problems facing the international community continue to unfold that create deep problems of justice, fairness, loss of life and destruction of highly valued natural resources, ethical principles which should guide environmental policy formation on these matters are failing to achieve traction even in cases where the ethical dimensions of the environmental controversy should be obvious. The chapter then explores reasons for the continuing lack of influence of ethical principles in environmental policy-making. Finally, the chapter makes recommendations on how to achieve greater traction for ethical principles in environmental decision-making.

Growth of the academic discipline of environmental ethics

The academic field of environmental ethics did not exist as a distinguishable discipline until the late 1970s. Although many writers explored ethical problems with human destruction of the natural environment before then, including Rachel Carson in *Silent Spring* in 1962 and Aldo Leopold in *A Sand County Almanac* in 1949, environmental ethics was not understood to be a distinguishable ethical subdiscipline until the end of the 1970s. With the creation of several academic journals such as *Environmental Ethics* in 1979 and the *Trumpeter Journal of Ecosophy* in 1983; the significant rise in published environmental ethics papers; and the creation of academic conferences that expressly focused on environmental ethical issues, the field of environmental ethics was well established by the mid-1980s. The emergence of this subdiscipline followed shortly after the rise of the modern environmental movement in the 1970s, which many observers have attributed to the increasing visibility in the 1960s of numerous serious environmental problems, such as the Santa Barbara oil spill, the Cuyahoga River in Ohio catching fire, and the growing awareness of the widespread and toxic threat of pesticides, due in large part to Rachel Carson's *Silent Spring*.

The focus of academic environmental ethics has largely been about the obligations of humans to protect non-human species and natural resources, despite the fact that environmental problems have threatened human health and well-being and thereby have invited policy responses based on ethical principles that were well established long before the rise of environmental ethics in the 1980s. These principles include deontological, utilitarian, virtue ethics, human rights and distributive justice based considerations. Yet, as we shall see, environmental policy-making often ignores even these long-established ethical principles. At the same time, much of the academic environmental ethics literature has infrequently focused on environmental policy-making controversies that often call for responses based on these well-established ethical norms because the subdiscipline's major focus has been examining human responsibilities to non-humans and the natural environment at a theoretical level while rarely being directly engaged with scientists, economists, and policy-makers in environmental policy formation.

Recent evidence of lack of traction for ethical principles in guiding environmental policy-making

The claim that ethical principles are rarely guiding environmental policy formation is strongly supported by the founder of the journal *Environmental Ethics*, Eugene Hargrove, who in 2003 published an essay, 'What's Wrong? Who is to Blame?' (Hargrove 2003). This essay invited reflection on why environmental ethics has not had more of an influence on environmental policy. Just three years later, Robert Frodeman, also in *Environmental Ethics*, in an article entitled 'The Policy Turn in Environmental Ethics', also reflected on the failure of environmental ethics to influence environmental policy-making while examining what can be done to make environmental ethics more influential in policy-making (Frodeman 2006).

Recent examples of significant failures of environmental policy-makers to identify and respond to important ethical issues embedded in environmental policy formulation controversies include the following:

1. For thirty-five years a climate policy debate has raged in the United States in which opponents of proposed climate policies have largely made two kinds of arguments: one, that climate change policies should be opposed because of uncertainty about the magnitude of human-induced climate change impacts; and two, that proposed climate change policies are unacceptable because they will impose excessive costs on national economies or specific industries (Brown 2017). These arguments raise profound ethical issues, yet the public debate responding to this opposition has focused not on ethical problems with these arguments, but on factual claims about scientific uncertainty and the magnitude of economic costs of the policies.
2. When President Donald Trump on 1 June 2017 announced he was withdrawing the United States from the 2015 Paris Agreement on the basis of

putting American economic interests first, few, if any, of the criticisms of his decision discussed in the press noted that the United States not only has economic interests, but also ethical and moral obligations to poor people and nations who are most at risk from climate change.

3. Nations continue to approach international climate change negotiations as if their economic interests alone are a legitimate guide for domestic climate change policy formation rather than their ethical responsibilities to those most vulnerable to climate change impacts. Yet climate change is obviously a civilization challenging ethical problem because:

    a. High emitting nations and individuals are putting mostly poor people and nations around the world at greatest risk of harm, people and nations that have done little to cause the problem.
    b. The harms to climate change's victims are not mere inconveniences but potentially catastrophic losses of life or damages to ecological systems on which life depends.
    c. Most of those most vulnerable to climate change's harshest impacts in poor countries can do little to protect themselves by petitioning their governments to protect them; their best hope is that high emitters will see that they have duties to the victims to lower their greenhouse gas (GHG) emissions.
    d. Any national GHG reduction target is implicitly a position on two ethical questions: (1) how much harm and death may the nation inflict on others without their consent; and (2) what is the nation's fair share of safe global GHG emissions.

4. When Pope Francis in May of 2015 issued his 'Laudata Si' encyclical which called climate change a moral issue, it got worldwide attention (Pope Francis 2015). Yet despite broad international media coverage of the Pope's claim that climate change is a moral issue, there has been virtually no media coverage of the ethical problems with the failure of governments to act to prevent damages from climate change.
5. Research conducted by Widener University Commonwealth Law School and the University of Auckland revealed that despite the fact that all nations agreed under the United Nations Framework Convention on Climate Change to establish GHG emissions reduction targets at levels necessary to prevent dangerous anthropogenic interference with the climate system at levels entailed by equity and common but differentiate responsibilities, most nations actually set their national GHG reduction target on national economic interest while ignoring equity and distributive justice (National Climate Justice 2016). Yet public response to national GHG targets has largely ignored numerous ethical problems with the national GHG reduction targets (ibid.).
6. An executive order of the United States president requires any US proposed regulation must satisfy a cost-benefit analysis before it may be promulgated (Congressional Research Service 2014). This is so despite

the fact that cost-benefit analysis used as a prescriptive guide to rulemaking often hides controversial ethical issues that very frequently arise in environmental policy-making, including the failure of nations to comply with their duty to not harm others on the basis of national self-interest (Brown 2010).

Causal explanations for ethics' lack of traction

Concerned observers of the failure of ethical considerations to guide environmental policy attribute the following causes to this phenomenon.

1 *The dominance of instrumental rationality in decision-making*

Some observers attribute the failure of ethics to obtain traction in environmental policy-making to a broader social reality, namely the dominance in the modern world of instrumental rationality to formulate public policy at the expense of ethics or moral reasoning. Cruickshank explains that:

> Instrumental rationality is a mode of rationality that is exclusively concerned with the search for efficient means and which, consequently, is not concerned with assessing the goals – or ends – pursued. This form of rationality has existed throughout history, but has become dominant in post-Enlightenment liberal democratic capitalist societies.
>
> (Cruickshank 2014)

Instrumental rationality, because it focuses on means, and not ethical ends, hides or ignores ethical questions about what the goals of policy should be. Ethical reasoning is concerned with what should be the goal of human behaviour including what is right or wrong, what behaviour is permissible or impermissible, and what actions are obligatory or non-obligatory. Scientific and economic reasoning, which has dominated public policy-making from the beginning of the Enlightenment, focuses on how to achieve goals, not on what goals or ends should be desired. Scientific reasoning based upon empiricism tests hypotheses to determine what is. Most moral philosophers believe that determining what 'is' is the proper domain of science. Yet science cannot determine what 'ought' to be which is the domain of ethics. Yet the instrumental rationality that scientists and economists deploy in their search for scientific and economic facts has dominated public life and higher education for several centuries despite the fact that environmental policy-making raises numerous ethical issues.

That instrumental rationality dominates environmental policy-making is clear given that government environmental agencies are staffed almost exclusively by lawyers, engineers, scientists, and occasionally economists, but very few by employees trained in ethics.

Economic reasoning is often exclusively based on instrumental rationality because economists most often assume that economic growth and efficiency should be the goals of public policy. Given this starting assumption, most economic analyses of proposed environmental policy options analyse societal costs and benefits of proposed environmental policies or determine how to structure policies to achieve maximum economic efficiency. Thus, the economic analyses of proposed environmental policies usually examine how to achieve economic goals, where the ultimate goals of policies are not in question. Economists may be able to determine which human preferences will be most satisfied by responses to public policy options, but they cannot determine what preferences humans should have because this is the domain of ethics, not economics.

Scientists employed by environmental agencies usually focus on understanding the environmental harms and risks of various human activities and whether proposed government action will acceptably reduce threats to human health and the environment. The goals of environmental regulatory action are usually given to them by law or regulation such as water pollution should be reduced to prevent unreasonable harm to humans or ecological systems. In the face of scientific uncertainty about whether human actions may cause great harm, scientists cannot determine who should have the burden of proof or what quantity of proof should satisfy the burden of proof by scientific methods alone because these are fundamentally ethical questions.

2 *Opponents of environmental policies frame arguments against proposed policies in a way that ignores ethical issues*

Opponents of environmental policies almost always object to proposed policies on the grounds that the policies will impose unacceptable costs or that there is insufficient scientific proof to justify regulatory action. They thereby trick proponents of environmental policies to argue about the magnitude of regulatory costs and scientific uncertainty while ignoring ethical issues entailed by environmental problems, such as whether a person or entity has a duty not harm others on the basis of economic self-interest, or the duty to refrain from dangerous behaviours that may harm others. Thus, ethical questions entailed by environmental policy formulation controversies remain unexamined. Most employees of government environmental policy offices are technically trained in science or economics; therefore, they are often poor in spotting ethical problems with arguments made about policy.

3 *The failure of higher education to educate policy-makers about ethical questions*

Higher education is largely failing to train those who will be engaged in environmental issues to spot ethical issues. Although many schools teach environmental ethics usually as an elective, most students enrolled in environmental economics, science, or engineering do not have any ethical training. And so, most of the technical professionals working for environmental agencies have no training in ethics that would help them spot ethical issues. Higher education usually trains students in economics and environmental science without any exposure to the ethical questions that frequently arise in the application of these disciplines to environmental policy. For instance, students in environmental studies programs are frequently taught to analyse environmental issues by using economic analytical tools such as cost–benefit analysis (CBA). These analytical economic tools are also often taught as if they are 'value-free' and, as a result, students are simply taught the analytical techniques for determining such things as the economic value of costs and benefits and how to calculate these values in performing cost-benefit analyses. Students are rarely taught the ethical limits of these tools when they are applied to public policy controversies such as climate change. Because economics and science are technical disciplines that require training in higher education to understand and decipher the application of these disciplines to policy formation, most citizens fail to spot ethical questions entailed by the scientific and economic analyses that frame most environmental policy challenges, such as climate change. In other words, civil society is usually not equipped to spot ethical issues that arise in scientific and economic descriptions of technically complex social and environmental problems.

4 *The narrow scope of academic environmental ethics training*

University education on environmental ethics frequently fails to cover ethical questions that arise when environmental economics and science are applied as prescriptive guides to public policy. As discussed above, this is so because the major focus of academic environmental ethics has been to explore ethical questions about human duties to non-human nature, not ethical questions that arise in policy formation, such as the imbedded normative assumptions of economic arguments about proposed policy, problems of procedural and distributive justice, or the ethical issues that arise when government officials must make decisions in the face of uncertainty.

5 *Lack of engagement of environmental ethicists in policy formation processes*

Academics working in the discipline of environmental ethics rarely engage in policy-making by working with scientists, engineers, and economists who dominate environmental agencies to help the government spot and respond to many of the ethical questions raised by policy controversies. Their academic focus has usually been examining theoretical questions about human duties to protect humans or natural resources from threatening human activities. Thus, many academic environmental ethicists have rarely been involved in applying ethics to actual environmental controversies when they are considered in policy formation.

6 *Much environmental ethics literature is not relevant to the ethical issues that arise in policy formation*

Because certain value-neutral policy languages structure specific environmental controversies, and because the environmental ethics literature does not usually focus on ethical analyses of concrete problems, most environmental ethics literature is not relevant to some of the most frequent issues that arise in policy-making, namely economic and scientific arguments about whether to act to protect the environment. As Frodeman has explained, environmental ethics research has rarely engaged in environmental policy formation working with scientists, engineers, and economists who are responsible for formulating environmental policy in environmental agencies (Frodeman 2006). A strong case can be made that the field of environmental ethics that bloomed in the late 1970s was based, at least in part, upon a misdiagnosis of the causes of ethical failure to protect the environment. In the beginning of the subdiscipline of environmental ethics, the environmental ethics literature often assumed that a narrow anthropocentric ethical footing that dominated twentieth century Western culture was the primary cause of the emerging environmental crisis. As a result, environmental ethics literature has primarily consisted of ethicists responding to each other about how to reform, replace, or extend dominant anthropocentric ethical norms about environmental problems. Yet, environmental ethics could make an important contribution to environmental policy-making by helping citizens understand the ethical problems with instrumental rationality that very frequently structures scientific and economic arguments about environmental policy. Moreover, because many, if not most, ethical issues that arise in environmental policy formation can be analysed through the lens of non-controversial ethical theories such as duties to do no harm, the responsibility of governments to secure human rights, and the requirement of procedural justice to secure the consent of those who will be harmed by failing to act, much of the theoretical environmental ethics literature that has been the primary focus of academic environmental ethics is not relevant to actual environmental controversies.

7 *Advocates of environmental policies usually fail to spot ethical problems with the opponents of proposed environmental policies*

Not only do opponents of environmental policies frame policy arguments as if they were 'value-neutral' facts, advocates of environmental action also often fail to identify the ethical problems with arguments made against environmental policies. For instance, when opponents of climate change policies claim that climate change legislation should be opposed because it will cost the national economy too much, advocates of environmental policy responses often simply reply by pointing to potential increases in green jobs. What goes unnoticed in this debate is that nations and citizens have duties to people around the world to prevent harm, and given that GHG emissions are already and will in the future be the cause of great suffering to people and ecological systems, high-emitting nations and entities have a duty to reduce GHG emissions to all who will be harmed, and economic self-interest alone is a deeply ethically problematic excuse for not taking action to prevent harm. When opponents of climate change policies claim that there is too much uncertainty to warrant government reductions of GHG emissions, advocates of climate change policies have often responded by claiming that climate science is settled, while ignoring ethical issues entailed by decision-making in the face of scientific uncertainty, such as who should have the burden of proof and what quantify of proof should satisfy the burden of proof.

8 *The media fails to cover ethical problems with arguments of the opponents of environmental policy*

Because environmental professionals and policy-makers don't spot the ethical questions, the media almost always fails to cover the ethical issues entailed by the economic and scientific arguments that have been the central discourses in environmental policy controversies. As a result, there has hardly been a peep or a whimper in the press about the ethical issues entailed by climate change policy controversies, despite the fact that some environmental controversies such as climate change raise deep and obvious civilization-challenging ethical questions.

9 *Specific ethical problems with economic arguments made against proposed environmental policies are ignored*

Economic arguments against environmental problems often hide problematic ethical issues. If, for instance, a CBA concludes that regulatory action to protect vulnerable people or ecological systems should not be taken because costs of taking action to reduce an environmental threat outweigh the economic value of harms avoided by the proposed regulation, controversial ethical assumptions may be hidden in factual assertions about the value of benefits if:

a. Potentially but not fully proven catastrophic harms were ignored in the CBA.
b. The costs of taking action would be imposed upon parties that are harming others, yet the victims of the harm have not consented to be harmed.
c. Things that are believed to be sacred by one culture are valued in the CBA as if they were commodities whose value can be measured exclusively by 'willingness-to-pay' monetary measures.
d. CBA commodify all human values usually only by considering monetary value and ignoring sacred value or ethical considerations for concluding that certain entities should not be for sale.
e. Human rights will be violated if regulatory action is not taken.
f. The proposed regulatory action implements the ethical duty to not harm others.

g. CBA determined the economic value of entities that might be harmed without obtaining the consent of those who might be harmed by the environmental threat.
h. The benefits of government action to protect the environment are discounted too greatly to allow future values that will be experienced by others to be compared with current costs of those responsible for environmental threats.

10 *Specific ethical problems with scientific uncertainty arguments made against proposed environmental policies are ignored*

If a government decides not to act to reduce the threat of environmental harm on the basis of lack of proof of harm, such a decision can hide important ethical questions if:

a. The government assumes that the proponents of climate change action should shoulder the burden of proof of demonstrating harm even in matters where proof is expensive or difficult to demonstrate.
b. There is credible but uncertain evidence that the current activity may be approaching thresholds that could trigger serious consequences.
c. The government waits until all uncertainties are resolved it will be too late to prevent serious harm.
d. The victims of the harm have not consented to be put at risk by a decision to not act in the face of uncertainty.
e. Some very serious potential harm is judged to be low probability just because the mechanism for causing the harm is not completely understood so that the probability of the serious harm cannot be confidently evaluated.

11 *Scientists and economists working for government in policy positions often assume that their job is solely to determine scientific and economic facts*

Government decision-makers usually assume that their responsibility is to apply technically determined 'facts' derived from scientific and economic analysis to policy guidance contained in relevant law. Although it is true that technicians working for government are expected to defer to legislative guidance for policy guidance, and government decision-makers are not supposed to decide cases based upon their own values, but to apply facts to policy rules decided in a democracy by law, this common understanding of government responsibility mistakenly assumes that science and economics can derive 'value-neutral' facts about environmental controversies. Yet, making sense of scientific and economic 'facts' about environmental problems frequently requires value-judgments about which economic 'facts' count, how environmental entities should be 'valued' in economic descriptions of the consequences of environmental policy, what levels of scientific risks are acceptable, what to do about potentially catastrophic but low-probability harms, and how much money should be spent in environmental investigations to better understand potential causes of environmental problems. There are simply no 'value-neutral' ways of deciding these more contentious questions about how to apply scientific and economic conclusions to legislative guidance on many environmental controversies.

Recommendations

To achieve greater consideration of ethics as a guide to environmental policy-making, certain strategies should be deployed. They include:

1. Government agencies that formulate environmental policy, and scientific organizations which make environmental policy recommendations, should establish an office, organizational entity, or person to engage in ethical analyses of proposed environmental policies.
2. Environmental ethicists must engage with scientists, economists, and policy-makers on issues that arise in policy formation, including issues raised by arguments of both proponents and opponents of environmental policies.
3. Environmental ethicists must give greater research focus to applied environmental ethical questions while seeking to assure that research results are completed timely enough to be considered by policy-makers in policy formation processes.
4. Because different valid ethical theories may not always agree on what ethics requires, ethicists must help policy-makers understand:

    a. when there is an agreement among ethicists on what ethics requires;
    b. when there is a disagreement among ethicists on what ethics requires and the different conclusions of applied ethical theories;
    c. the reasons for favouring one ethical approach among others;
    d. the need to display how ethical controversies were resolved to allow public reflection on these matters; and
    e. processes for resolving ethical disputes, which includes an opportunity for participation by all stakeholders in a deliberative process.

5. Policy-makers should invite public participation on resolving contentious ethical questions for important policy questions. Policy-makers need to raise the visibility of ethically problematic positions of participants in policy debates.
6. A policy turn in academic environmental ethics is greatly needed to give attention not only to researching theoretical ethical questions, but to find ways of engaging with scientists, economists, and policy-makers in environmental policy formation.
7. Environmental ethicists need to find ways of educating policy-makers and citizens about the ethical issues that are hidden in what appear to be value-neutral arguments about scientific and economic issues in environmental policy formation.

8. Higher education needs to assure that all environmental education includes training on applied environmental ethics with a particular goal of exposing students in policy sciences to the role of ethical issues that arise in policy-making.

References

Brown, D., 2010, Ethical Problems with Cost Arguments against Climate Change Policies: Increased Cost May Not Justify Human Rights Violations, available at https://ethicsandclimate.org/2010/09/20/ethical_problems_with_cost_arguments_against_climate_change_policies_increased_costs_may_not_justify.

Brown, D., 2017, The Moral Outrageousness of Trump's Decision on the Paris Agreement, Ethics and Climate, available at https://ethicsandclimate.org/2017/07/07/the-moral-outrageousness-of-trumps-decision-on-the-paris-agreement.

Carson, R., 1962, *Silent Spring*. Boston, MA: Houghton Mifflin.

Congressional Research Service, 2014, Cost–Benefit and Other Analysis Requirements in the Rulemaking Process, available at https://fas.org/sgp/crs/misc/R41974.pdf.

Cruickshank, J., 2014, Democracy versus the Domination of Instrumental Rationality: Defending Dewey's Argument for Democracy as an Ethical Way of Life, *Humanities*, 2014, 3, 19–41.

Frodeman, R. 2006 The Policy Turn in Environmental Philosophy, *Environmental Ethics*, 28, 3–20.

Hargrove, E. 2003, What's Wrong, Who Is to Blame?, *Environmental Ethics*, 25(1): 3–4.

Leopold, A., 1949, *A Sand County Almanac*. Oxford: Oxford University Press.

National Climate Justice, 2016, National Climate Justice, A project of Widener University Commonwealth Law School/University of Auckland, available at https://nationalclimatejustice.org.

Pope Francis, 2015, Laudato Si, On Care for Our Common Home, available at http://w2.vatican.va/content/francesco/en/encyclicals/documents/papa-francesco_20150524_enciclica-laudato-si.html.

16 Planet ocean and marine protected areas

An opportunity for ecological commons governance

Prue Taylor

Introduction

As land dwellers, we frequently overlook that we live on Planet Ocean. Comprising around 70 per cent of the Earth's surface and representing 95 per cent of the living biosphere, the global ocean is one of the greatest ecological commons on the planet. It is the source and sustain*er* of all life, unifying humanity in a manner more tangible than the global atmosphere. In contempt of this, the ecological integrity of the global ocean (comprising five named oceans) is in a perilous state. To address this, humanity must confront one of the largest ecological law and governance voids on Earth.

This chapter considers this void from the perspective of marine protected areas (MPAs) as a tool for protection of ecological integrity and their use in a new legal regime for 'conservation and sustainable use of marine biological diversity'. It argues that far from filling the void, this regime represents a 'retreat' from normative development of oceans law (at worst) or little more than 'business as usual' (at best). Employing an ecological law and governance approach, it considers how the 'common heritage of mankind' could be used to protect and restore the ecological integrity of the global ocean. It concludes by recommending that current state discussions be suspended and replaced by an inclusive process of common*ing*, at a global level. Without this it seems unlikely that humanity will move from state/market governance to commons-based governance, based on a normative principle of human responsibility to protect and restore the ecological integrity of the oceans, for the benefit of all.

Current context

In recent decades, humanity has developed a more comprehensive understanding of the global ocean as a complex interconnected ecological system and an integral part of the Earth's climate system. At a higher level, the oceans are part of one complex Earth System (Steffen 2016). Earth system science establishes what we have long known: the oceans are both the source and the *sustainer* of all life on Earth. Thus the oceans belong to all life (past, present and future) in the most elemental sense. For this, reason many cultures traditionally refer to the oceans as our collective Mother.

Despite the profound significance of the oceans for all life on Earth, their ecological integrity is in a perilous state. The *cumulative* impact of multiple human activities has caused massive degradation and dangerous tipping points. In addition, human demands are increasing in scope (new uses) and magnitude (scale of use). Once considered 'boundless and bountiful', we are awakening to the reality that their regenerative capacity and ecosystem function face profound challenges (World Oceans Assessment 2016). After years of relative neglect, the plight of the oceans is now receiving more attention. The United Nations (UN) initiated a World Oceans Assessment on the state of the marine environment, including social and economic aspects (ibid.). Sustainable Development Goal (SDG) 14 recognizes the centrality of oceans to a sustainable development agenda. As regards climate change, the oceans are mentioned in the 2015 Paris Agreement and will be the subject of an Intergovernmental Panel on Climate Change special report.[1] While this activity is welcome, as experience suggests, more information alone will not guarantee effective responses.

Over the last few decades, many responses have emerged to protect the oceans. Most reflect traditional resource conservation approaches; focusing on maintaining supply of natural 'resources' for human consumption. However, MPAs have emerged as a tool to protect and restore marine ecosystems. Defined as areas under special management to 'conserve the biological diversity and productivity (including ecological life systems) of the oceans' (IUCN 2012), the extent to which human activity is allowed varies. In 'reserves' (a type of MPA) activity is usually strictly limited. In other MPAs, some activity consistent with MPA goals is allowed. The ecological benefits (especially in fully protected reserves), include; protection of biodiversity, ecological benefits for neighbouring ecosystems, predator protection and creation of scientific reference points through removal of some ecosystem stressors. However, MPA design and good management of is critical to achieving ecological benefits. They work best if they are large, well-managed and resourced, isolated (yet networked with other MPAs) and long-lasting. In the absence of these factors (and others), outcomes are variable (Pew Brief 2016). Furthermore, MPAs (on their own) create a dichotomy between special areas (worth saving) and non-special areas (which are not). If we are to protect and restore the marine environment as an integrated whole, then this dichotomy (when used alone) is untenable (Bastmeijer, 2016).

Despite global recognition of the need for, and benefits of MPAs, very few exist. Only 2.8 per cent of the global ocean is protected, with most MPAs occurring within territorial seas (9.7%). Only 0.14 per cent of areas beyond national jurisdiction (i.e. High Seas) are protected.[2] A current scientific assessment suggests that 30 per cent of the world's oceans need protection by MPAs, including reserves (O'Leary 2016). The 2014 World Parks Congress called for protection of 30 per cent of each ocean habitat and 30 per cent of the world's oceans, overall (Recommendations 2014). In stark contrast, states have only agreed to 'conserve at least 10 per cent of coastal *and* marine areas by 2020' (SDG 14).[3] However, to achieve either of these targets, humanity must resolve one of the largest ecological law and governance voids on Earth. As Judge Weeramantry observed, international environmental law remains a reflection of sovereign state interests, 'unrelated to the global concerns of humanity as a whole' (Weeramantry 1997). As a result, states do not accept legal responsibility to act in the 'greater interests of humanity and planetary welfare' (ibid.). There is no global legal regime for the creation, management and enforcement of MPAs *and* no means to require such a regime. The global ocean (one of the planet's greatest ecological commons) is largely unprotected at a time when its regenerative capacity is imperilled.

As will be outlined below, the ocean's ecological integrity is at the mercy of a law and governance system characterized by the tragedy of fragmentation *and* the tragedy of the commons, together with enclosure. A new legal regime for 'conservation and sustainable use of marine biological diversity' will likely perpetuate these problems.

Ocean law and governance

Current ocean law and governance reflects state sovereignty as the supreme value (Treves 2010). This has resulted in a highly complex and fragmented system of legal jurisdictions that create an ecological nonsense. One of the main legal divides is between areas within national jurisdiction (over which state sovereign interests dominate) and 'areas beyond national jurisdiction' (ABNJ). State jurisdiction over 'exclusive economic zones' (EEZs) and 'continental shelves' (CSs) grants significant use rights and very few counterbalancing duties of environmental protection. Furthermore, states are adept at progressively increasing the size of these zones through a process of 'creeping jurisdiction'. This results in a gradual enclosure of the global commons. The dominance of state sovereignty over these zones leaves vast parts of the oceans vulnerable to national interests and the efficacy of national environmental law (if any). Unsurprisingly, outcomes range from bad to mediocre, given differing interests and capacities of states. Outcomes can be further undermined by a lack of co-ordination with the management of human activities in other EEZs and ABNJ. In short, states are not under any effective legal duty (beyond that created by their own legal system) to protect the ecological integrity of the oceans *within* their jurisdiction.

The legal regime for ABNJ (or High Seas) poses even greater risks to ecological integrity. The dominant principle is 'freedom of the high seas'. This creates an open access regime for around 198 million square kilometres of the oceans (Lodge 2014). What began as a doctrine suited to the needs of maritime empires, has become a right to overfish and a license to pollute – i.e. the tragedy of the commons (Beesley 2004). Exercise of these freedoms or rights (e.g. to fish, navigate, lay pipes and cables) are only weakly constrained by the rule of 'due regard' to the interests of other states exercising those same rights (Law of the Sea Convention, Art. 87(2)).

Supplementing the 'freedom of the high seas' are two other international legal regimes of limited scope. Both have a human resource use and conservation focus, providing for some measure of environmental protection.

The first is the 1995 Fish Stocks Agreement which only applies to a small set of highly migratory and straddling fish species. The Agreement limits freedom of the High Seas, for the purpose these species (and associated) environmental protection. The second is the Area regime created by the 1982 Law of the Sea Convention (hereafter LOSC) under which the seabed, ocean floor and subsoil (together with its mineral resources), i.e. the Area (LOSC, Art. 136), are the 'common heritage of mankind' (CHM). This replaces freedom of the high seas with a concept of commons management (for the Area) for the benefit of all humanity, including future generations. Judicial comment confirms that CHM places a high level of environmental responsibility on states (and others), when conducting mining activities in the Area, for the benefit of all (ITLOS 2011). This is reflected in the International Seabed Authority's (ISA) environmental regulations. These aspects of CHM are consistent with academic opinion; states are required to act as environmental trustees for all humanity (e.g. Wolfrum 2008). However, this obligation only applies to activities in the Area despite initial efforts to apply CHM to *all ocean space as an ecological whole* (Pardo 1993). Furthermore, its real impact will only be tested once commercial mining begins. For example, will a weak 'balancing of interests' approach be used enabling the ISA to sacrifice (or 'trade off') ecological integrity for economic benefit to states?

More generally, LOSC contains three environmental protection obligations that apply to the *marine environment as a whole* (LOSC, Art. 192 193 194). These duties have not delivered meaningful ecological outcomes because they have been applied within a context of 'weak' international co-operation (where individual and collective state interests are the priority). There is general agreement that the ability of other treaties (e.g. 1992 UN Biodiversity Convention) to provide for the protection of ecological integrity, is very limited especially for the High Seas.

Overall, because state sovereignty is the supreme value of oceans' governance, the underlying dynamics are of creeping jurisdiction (on one side) and open or unrestricted access (on the other). In both instances, the oceans are a legal object of human use and exploitation. This facilitates the current model of state/market based governance that frames the 'public interest' as a range of economic values best met through the actions of private corporate entities (Quilligan 2012).

In light of the critical state of the oceans, together with the ecological law and governance void described, the question is: will states respond by creating a regime with responsibility for ecological integrity at its core? Or will the response be 'business as usual'?

Current trends

States have been discussing oceans governance for ABNJ since 2002. From 2015 onwards, preparatory committee work has been framed as a regime for the 'conservation and sustainable use of marine biological diversity'. Full treaty negotiations on an implementing agreement to LOSC are now expected to begin in 2018, following a recent UNGA resolution (A/RES/72/249).

A critical element of this new regime is that it applies *only to ABNJ*. This is so despite the ecological nonsense it perpetuates, including the need for co-ordination between the management of areas within state jurisdiction *and* ABNJ. Furthermore, until recently, the primary focus has been on a regime for use and benefit sharing of marine genetic resources (MGR), reflecting the priority given to economic interests. In stark contrast, civil society's advocacy has focused on protection of *all* marine biodiversity, in addition to MGR. To date, their concerns have been narrowed down to a separate negotiation stream for 'area based conservation measures' including MPAs, but only in ABNJ. There has been little discussion of specific MPA targets (e.g. 10% or 30%) and how they will restrict traditional freedoms. More recently, a pragmatic package approach has been adopted meaning progress on MGR is conditional to progress on conservation measures (including MPAs) (Ardron et al. 2014). The acronym used for both streams is now BBNJ referring to marine 'biological diversity beyond national jurisdiction' which might suggest equal weight or a 'greenwash'.

As regards normative principles, there is a troubling lack of agreement. Some powerful states (e.g. US and Canada), favour 'freedom of the high seas'. China, G77 countries and small island developing states are generally supportive of CHM. However, in most instances CHM is being debated as a normative principle for MGR only, with very little discussion of its relevance to MPAs (ENB 2017). This reflects an erroneous understanding of CHM as a normative principle that *only* applies to regimes for resource utilization in ABNJ which (for equity reasons), justifies elements of capacity building and benefit sharing between states. As a result, ecological trusteeship of states – the most essential element of CHM – is ignored or overlooked.

Strong contention over CHM versus freedom of the high seas resulted in the EU proposing a 'practical approach', meaning discussions focus on practical elements of a regime rather than on guiding principles (Millicay 2015). This may be a useful strategy to keep discussions moving or a very troubling 'retreat' from the normative development of LOSC (considered the *Constitution of the Oceans*).

The importance of agreed normative principles cannot be underestimated. From an ethical perspective, they articulate shared values, over-arching purposes/priorities and identify beneficiaries. Whether implicit or explicit, they will be the foundation of any negotiations. More specifically, they define the start point or 'default position' of a legal regime. This has been defined, in environmental law, as the start point for what is acceptable human activity, in the absence of specific law (Jóhannsdóttir 2009). It also provides a guiding concept for more specific law (e.g. in treaties and statutes) thereby defining the spirit or intention, according to which law (including specific rules) is written, interpreted and applied (and when necessary) amended (Taylor, 2014).

This retreat from normative development could occur in two ways. First, the legal text may omit or make vague open references to, applicable normative principles. For example, there may be reference to capacity building and benefit sharing of MGR, without specific mention of CHM (Millicay 2015). MPAs may be dealt with as a suite of specific rules for establishment, management and enforcement. Secondly, the text may explicitly reiterate the centrality of 'freedom of the high seas' (and pre-existing rights and duties under LOSC) and CHM (for the Area) but include some very narrowly defined limitations for MGR and MPAs.[4] Either way, the key point is this: *implicit* in these approaches is continued support for the pre-existing ethical position of oceans as resources for human use and exploitation. This facilitates and justifies further enclosure and dispossession of the commons through creeping jurisdiction and other forms of appropriation. Ocean grabbing can continue. A strong 'rights' focus is maintained, subject only to limited reciprocal duties. Some specific and limited constraints on traditional high seas freedoms may be achieved, but at the cost of further fragmentation and continued potential for tragedy of the commons, outside of MPAs and MGR. In short, a weak form of state co-operation for the purpose of securing state interests (individual and collective) persists.

Informed BBNJ commentators would not be surprised by this analysis. With sovereignty remaining the supreme value, states reject any limitations. Equally, many states are in no hurry to further restrict their 'freedoms of the high seas' perceiving their sovereign interests to be well served by this principle (Treves 2010). A compounding factor is the failure of civil society (and states) to clearly articulate an alternative normative concept capable to creating a new default position. Thus far, the only candidate offered by civil society is 'common concern of humanity', a relatively weak concept (Taylor 2018).

In conclusion, states have thus far retreated from normative development of oceans' governance due to entrenched interests and an abject failure to respond to the magnitude and urgency of threats to oceans and all life. The minimal international attention given to BBNJ is (perhaps) indicative of its lack of ambition! The next section outlines the normative basis for law and governance which puts responsibility for protection of the foundations of life, for the benefit of all, at its centre.

Ecological law and governance for the oceans

Ecological law and governance makes ecological integrity a precondition for human aspirations and a fundamental principle of law. It reverses the principle of human dominance over nature, to a principle of human responsibility for nature. Based on eco-centrism and intra- /intergenerational and interspecies justice, it recognizes ecological interdependencies and favours collective responsibilities over individual rights (Oslo Manifesto 2016).

Ecological law and governance is an effort to transform the current paradigm and ethical basis of traditional environmental law. CHM is a pre-existing legal concept consistent with this effort. When properly understood, beyond its historical treatment by states in LOSC and other treaties, CHM has ecological trusteeship of states at its core. CHM creates a legal regime according to which states become charged with legal responsibility to prioritize and act consistently with the common interests of all humanity. States are no longer free to act solely in *their* individual national or collective self-interests. This means that the acts and the omissions of states can be measured against a legal responsibility to protect and restore ecological systems, for the benefit of all (Taylor 2016). Kiss summarizes the trust elements in these terms:

> [S]tates suspend or do not assert rights or claims, or in some cases exercise such jurisdiction only within set limits, *for the benefit of the whole human community*, without any immediate return, and conserve and if necessary manage areas *in conformity with the common interest for the benefit of all mankind*.
>
> (Kiss 1985: 427; emphasis added)

With 'trust' at the core of CHM, its use is no longer restricted to the international management regimes of exploitable resources in ABNJ. The key is that states act on behalf of humanity and in its interests (ibid.). While CHM appears to be an anthropocentric concept by focusing on 'environmental' protection and the mankind's interests/benefits, instead of ecological integrity and the interests of all life, including humanity. On closer inspection, CHM embraces a moral force that unifies humanity and encompasses respect for each other *and* all living things (Mann Borgese 1986: 131; 1998: 198).

Using CHM as a normative principle for the oceans would require applying it to the whole global ocean as a unified ecological commons. This is a return to the original proposal for CHM, prior to LOSC (Pardo 1993).[5] When applied *within* national jurisdiction, states retain legal power to control and regulate activities. But sovereign authority is subject to limitations (ecological responsibilities) specifically designed to protect the interests of all. This creates a legal basis *requiring states* to (a) create and also maintain the necessary network of MPAs within their jurisdictions (determined by an expert body) *and* (b) manage all other human activities (e.g. fishing and mining) consistent with MPA objectives wherever located (e.g. in adjacent EEZ's or High Seas). Ultimately, ecological responsibilities for marine ecosystems should also influence relevant land use management. Applied to ABNJ, CHM requires states to (a) create the necessary network of MPAs (determined by an expert body)[6] *and* (b) manage all high seas freedoms *and* activities within their jurisdiction/control, consistent with MPA objectives wherever located (e.g. in adjacent EEZ's or High Seas). The onus is on the proponents of all new activities to demonstrate consistency with MPA objectives. Given the different capacities of states, very careful attention to benefit and burden sharing will be required both within jurisdictions and ABNJ.

Applied in the above manner, CHM creates a new default position (transforming state sovereignty and replacing freedom of the High Seas). The global ocean becomes the legal object of a commons governance regime characterized by the fulfilment of ecological responsibilities. The task of creating, maintaining and managing MPAs becomes a requirement for meeting those responsibilities, and not a management option employed at the pleasure of states when (and to the extent that) it meets a consensus view of individual and collective interests. Furthermore, CHM guards against distortion of key MPA elements. For example, their purpose is ecological not 'replenishment of fish stocks', or the protection of military activities. Critically, the nature, duration and location of activities (if any) allowed within MPAs, should all be determined by ecological objectives and not negotiated time limits (sunset clauses) and the sites of commercial fisheries.

As a new default position CHM would also play an important co-ordinating role for development of pre-existing legal regimes (e.g. for the Area and Fish Stocks Agreement) and any new specific regimes (MGR). The objective would be to ensure that all regimes evolve, develop and remain consistent with ecological protection as the priority.

A final critical element of CHM is non-appropriation. This is relevant to the use of CHM in ABNJ and would prevent states further claiming or exercising sovereignty/sovereign rights over ABNJ. The objective is to halt continued and new forms of creeping jurisdiction, thereby legally claiming and protecting ABNJ as a global ecological commons; belonging to all. This will require better delimitation of claims to national jurisdiction.

Conclusion

The above proposal is highly ambitious and beyond the scope of BBNJ and a LOSC implementing agreement, which adds to but does not amend the Convention. Furthermore, there is unlikely to be state support within the context of current negotiating processes dominated by powerful states and commercial interests. Therefore, rather than continuing it is suggested here that the BBNJ discussions be suspended and replaced by a highly inclusive process of common*ing*, at a global level.[7] The objective is collective acknowledgement and articulation of the global ocean as a commons, belonging to all. Once visible as a commons, states and civil society can (re)build our social/cultural and legal relationships. Central to this will be commitment to the oceans as the source and sustain*er* of all life and identification of values that unify, inspire and underpin the moral and legal responsibility of all. In essence, common*ing* will be central to building, what Ronald Engel defines as a 'covenant' (Engel 2004). Without this, it seems unlikely that humanity will fully commit (including with enforcement and sanction) to commons based governance, including a normative principle of human responsibility for the oceans ecological integrity, for the benefit of all.

Common*ing* will also be central to addressing the legitimacy issue. States claim exclusive legitimacy to create oceans law and governance, despite their obvious inability to protect ecological integrity. By requiring states to act as global ecological trustees, the legitimacy of states as actors in global governance, can be transformed and strengthened (Bosselmann 2015). Finally, common*ing* would reinforce the legitimacy of civil society representatives to sit alongside states as trustees – putting the *demos* back into public international law.

Notes

1 Due in 2019; see www.ipcc.ch/report/srocc.
2 See www.protectplanetocean.org/official_mpa_map.
3 See https://sustainabledevelopment.un.org/sdg14. This target originated as Aichi Biodiversity Target Number 11; see www.cbd.int/doc/strategic-plan/2011-2020/Aichi-Targets-EN.pdf.
4 The most recent report is not specific on how normative principles will apply. Rather a list of principles for further discussion is identified. (Report of the Preparatory Committee 2017).
5 An alternative proposal was for the whole of the High Seas to be the CHM.
6 Alternatively, the whole of the High Seas could be an MPA with integrated management zones.
7 This term is attributed to P. Lindebaugh (Bollier 2014).

References

Ardron, A. N. (2014) 'The Sustainable Use and Conservation of Biodiversity in ABNJ: What Can Be Achieved Using Existing International Agreements?', *Marine Policy*, vol 49, pp98–108.

Bastmeijer, K. (2016) 'Ecological Restoration in International Biodiversity Law: A Promising Strategy to Address our Failure to Prevent?', in M. Bowman, P. Davies and E. Goodwin (eds) *Research Handbook on Biodiversity Law*, Cheltenham: Edward Elgar

Beesley, A. (2004) 'Grotius and the New Law', *Ocean Yearbook*, vol 18, pp98–116.

Bollier, D. (2014) *Think Like a Commoner: A Short Introduction to the Life of the Commons*, Gabriola Island, BC: New Society Publishers.

Bosselmann, K. (2015) *Earth Governance: Trusteeship of the Global Commons*, Cheltenham: Edward Elgar.

ENB (2017) *Earth Negotiations Bulletin* (24 July), vol 25, no. 141, available at http://enb.iisd.org/vol25/enb25141e.html.

Engel, R. (2004) 'A Covenant Model of Global Ethics', *Worldviews: Global Religions, Culture and Ecology*, vol 8, pp29–46.

ITLOS (2011) 'Seabed Disputes Chamber of the International Tribunal for the Law of the Sea – Responsibilities and Obligations of States Sponsoring Persons and Entities with Respect to Activities in the Area: Advisory Opinion' (1 February).

IUCN (2012) 'Guidelines for Applying the IUCN Protected Area Management Categories to Marine Protected Areas', available at https://cmsdata.iucn.org/downloads/uicn_categoriesamp_eng.pdf

Jóhannsdóttir, A. (2009) *The Significance of Default.* (Uppsala Universitet), available at www.diva-portal.org/smash/get/diva2:173192/FULLTEXT01.pdf.

Kiss, A. (1985) 'The Common Heritage of Mankind: Utopia or Reality?', *Law in the International Community*, vol, 40, no. 3, pp423–441.

Lodge, M. (2014) 'Satya Nandan's Legacy for the CHM', in M. Lodge and M. H. Nordquist (eds), *Peaceful Order in the World's Oceans: Essays in Honor of Satya N. Nandan*, Leiden: Brill.

Mann Borgese, E. (1986) *The Future of the Oceans: A Report to the Club of Rome*, Montréal: Harvest House.

Mann Borgese, E. (1998) *The Oceanic Circle: Governing the Seas as a Global Resource*, Tokyo: UNU Press.

Millicay, M. F. (2015) 'The Common Heritage of Mankind: 21st Century Challenges of a Revolutionary Concept', in de Castillo (ed.), *Law of the Sea, from Grotius to the International Tribunal on the Law of the Sea*, Leiden: Brill/Nijhoff.

O'Leary, B. (2016) 'Effective Coverage Targets for Ocean Protection', *Conservation Letters*, doi:10.111/conl.12247.

Oslo Manifesto (2016) Oslo Manifesto, available at www.harmonywithnatureun.org/content/documents/20170404OSLO_MANIFESTO.pdf.

Pardo, A. (1993) 'The Origins of the 1967 Malta Initiative', *International Insights*, vol 9, no 2, pp65–69.

Pew Brief (2016) 'Marine Protected Areas Beyond National Jurisdiction', available at www.pewtrusts.org/en/research-and-analysis/issue-briefs/2016/03/marine-protected-areas-beyond-national-jurisdiction.

Quilligan, J. (2012) 'Why Distinguish Common Goods from Public Goods', in S. Helfrich and D. Bollier (eds), *The Wealth of the Commons: Beyond the Market and State*, Levellers Press.

Recommendations (2014) 'Recommendations of the Marine Cross-Cutting Theme at the 6th IUCN World Parks Congress in Sydney, Australia', available at http://worldparks-congress.org/downloads/approaches/ThemeM.pdf.

Report of the Preparatory Committee (2017) 'Development of an Internationally Legally Binding Instrument under the United Nations Convention on the Law of the Sea on the Conservation and Sustainable Use of Marine Biological Diversity of Areas beyond National Jurisdiction', available at www.un.org/depts/los/biodiversity/prepcom_files/Procedural_report_of_BBNJ_PrepCom.pdf.

Steffen, W. (2016) 'The Planetary Boundaries Framework: Defining a Safe Operating Space for Humanity', in P. Magalhaes (eds), *The Safe Operating Space Treaty: A New Approach to Managing the Earth's System Use*, Newcastle upon Tyne: Cambridge Scholars Publishing.

Taylor, P. (2014) 'The Earth Charter, the Commons and the Common Heritage of Mankind Principle', in L. Westra and M. Vilela *The Earth Charter, Ecological Integrity and Social Movements*, London: Routledge.

Taylor, P. (2016) 'The Concept of the Common Heritage of Mankind', in D. Fisher, *Research Handbook on Fundamental Concepts of Environmental Law*, Cheltenham: Edward Elgar.

Taylor, P. (2018) 'Common Heritage of Mankind and Common Concern of Humankind', in E. Orlando (eds), *Encyclopedia of Environmental Law: Principles of Environmental Law*, Cheltenham: Edward Elgar.

Treves, T. (2010) 'Principles and Objectives of the Legal Regime Governing Areas Beyond National Jurisdiction', in E. J. Molenaar and A. G. Oude Elferink (eds), *The International Legal Region of Areas Beyond National Jurisdiction: Current and Future Developments*, Leiden: Martinus Nijhoff.

Weeramantry (1997) Case Concerning the Gabcikovo-Nagymaros Project (*Hungary v. Slovakia*), [1997] *I.C.J. Rep. 7, separate opinion of Vice-President Weeramantry, at C(c).*

Wolfrum, R. (2008) 'Common Heritage of Mankind', available at www.mpepil.com.

World Oceans Assessment (2016) 'United Nations First Global Integrated Marine Assessment', available at www.un.org/depts/los/global_reporting/WOA_RPROC/WOA Compilation.pdf.

Part IV

The future of ecological integrity

17 Towards world federalism for international peace and a sustainable environment

Peter Venton

Introduction

Today the solution to the biggest problems facing *Homo sapiens* is a world federation for international peace and sustainable development based on the principles of federal democracy that apply to self-sufficient nation states. In view of the anticipated resistance of many nation states to give up sovereignty over these areas, a more likely route to realizing world federalism would be through initiatives of the United States of America (America). This view is based on the premises that America is the most economically and militarily powerful state in the world and that it is a federal democracy.

However, the reality of America is that it is neither a democracy nor a nation state but rather a de facto oligarchy that governs an economic empire. Although it has the institutional machinery of a democratic federal republic, the dominant culture of many America citizens is not democratic. Aristotle's ideal democratic citizen sought a moderate income for the purpose of giving him the leisure 'for pursuing man's two chief ends, politics and philosophy' (Newell 2011: 17). By contrast, the aim of a majority of American citizens appears to be to work hard for all the wealth they can get and use their leisure for the pursuit of happiness rather than the responsibility for participating in politics. In this light, proposals are made for formal lifetime learning for participation in their democracy by all Americans, an idea inspired by eighteenth-century philosopher John Lock's 'prescription of education to develop rational, enlightened citizens for peace in the common wealth'. (Sikkenga 2015: 82–83). As well, proposals are made for formal education of prospective and incumbent oligarchic leaders in ethics and competencies required for their main tasks in legislation and oversight management of the economy for the common good.

Finally three proposals are made for America to initiate a process for making the United Nations a truly world federalist institution for peace and stewardship of the world's commons. First Americans should replace its present neoliberal capitalist system with a democratic form of capitalism that features less economic inequality and slower economic growth. Secondly the American government should enter into negotiations with the Russian and Chinese empires to persuade them to replace their neoliberal capitalist systems for slower economic growth in the interest of preserving their environments. Thirdly it is proposed that America negotiate with Russia and China to downsize their military resources and operations along with those of America and place the remainder under the auspices of the United Nations as a world federalist institution for peace keeping.

From the American perspective the foregoing may appear utopian. However the chapter ends with a brief outline that suggests America in on the verge of a new political revolution that may lead to a more democratic economy and a foreign policy more supportive of United Nations governance for international peace and sustainable development.

Problems for *Homo sapiens*

At the beginning of the third millennium, the major problems that confront *Homo sapiens* are climate change, population levels in excess of the ecological limits of the planet, industrial capitalism, globalization and the continued presence of war and, in particular, the threat of nuclear war.

The problem of climate change is the challenge of limiting the increase in the average world temperature since the dawn of the industrial revolution to 2 degrees. If that target is missed and we slip to 3 degrees for example, *Homo sapiens* may be endangered in a century or two.

Globalization is a phenomenon of four interrelated elements: increasing international trade; liberalized investment (i.e. an increase in the flow of financial capital between countries); a substantial degree of human migration between countries and a substantial degree of exchange of ideas and information between citizens and governments of countries. The first two elements of international trade and liberalized investment represent a 'package deal for cheap labour and more greenhouse gas emissions' (Klein 2014: 93). In the globalization of the nineteenth century, 60 million Europeans emigrated between 1820 and 1920, three fifths to the United States, and many others within Europe and a significant percentage to Latin American countries (Wikipedia 2013). By contrast the late twentieth-century globalization allows very limited scope for emigration to other countries as borders are largely closed.

Industrial capitalism that began around the late eighteenth century is a mode of production of goods and services and their distribution that features four interrelated institutions: capitalist firms, market exchange, a monetary system based on bank credit and a distinctive role for the state in relation to the first three features. Capitalism comes in a variety of forms depending upon the role of the state in supporting or regulating the first three institutions (Mann 2013: 13). Problems of forms of industrial capitalism in most countries are that they are undemocratic and they produce excessive economic growth rates that destroy the earth's environment.

Solutions

The solutions to these problems are world federalism for international peace and security and stewardship of the world's commons, a change in the form of the capitalist economic system to render it democratic and reduce its excessive rate of economic growth and the introduction of policies for inducing a reduction in human population. This chapter deals with the first two solutions.

Presently world federalism for international peace and security is found in the United Nations Security Council (UNSC). It is one of the six principal organs of the United Nations charged with the maintenance of international peace and security. Its powers include the establishment of peacekeeping operations, the establishment of international sanctions, and the authorization of military action through Security Council resolutions. It is the only UN body with the authority to issue binding resolutions to member states. Today the five permanent members of the 15-member security council include the United States, Russia, The Peoples Republic of China, the United Kingdom and France. However, the permanent members have a veto which has been effectively used to neuter the UN mandate to govern for world peace (Wikipedia 2017a).

The problem with world federalism is that it has, for centuries, been regarded as utopian. In the seventeenth century, Thomas Hobbes argued that nation states would never feel the need to give up their sovereignty to form a world government authority (Johnson 2015). In the eighteenth century, Immanuel Kant rejected world federalism on grounds that extending dominion over the whole world would more likely induce tyranny than world peace (Bradshaw 2015: 135). Another problem is with the United Nations' concept of universal human rights – that is rights that apply to all and specifically to different races, ethnic groups, and national cultures, socio economic classes. By definition this implies that the desire for peace represents the will of all or at least a majority of the world's citizens. Yet not all nation states are democratic and therefore their governments may not support their citizens' rights to peace. Insofar as the world federalist movement seeks to apply the principles of democratic federalism to more effectively address global problems, the presence of democratic states is a critical prerequisite.

In light of these objections I believe that the route to world federalism might more likely be initiated by America on grounds that America is a well-developed federal democracy with the machinery that gives it the potential to peacefully change its present form of capitalism. This belief is also based on the observation that the present *de facto* alternative to international peace and security appears to be '*pax Americana* – a concept of relative peace in the Western Hemisphere and later the world as a result of the preponderance of power enjoyed by the United States beginning around the middle of the 20th century and continuing to this day' (Wikipedia 2017b). The high point of this concept was perhaps 1990 when the Soviet Union collapsed and both Russia and China had abandoned communism and adopted market capitalism. The other premise is that a majority of Americans regard world peace as their right and are prepared to insist that their government support world governance through the United Nations. A related premise is that they would benefit from a substantial peace dividend that they would receive as a result of a United Nations peace keeping authority. Finally, it is based on the premise that America's preponderance of economic power in the world would enable it to influence its rival empires of Russia and China on the security council to change their economic systems and submit to a world authority for international peace under the auspices of the United Nations. A caveat about this argument is the fact that the last twenty-five years have seen a revival of threats from Russia and China together with evidence of a substantial decline in the capacity of America to act as the world's police man.

Democracy in America

> In the most influential book on the founding published in the second half of the twentieth century, *The Creation of the American Republic 1776–1787*, Gordon Wood demonstrated convincingly through an exhaustive examination of the evidence that the architects of the United States were animated by a passion for civic virtue as well as liberty. They not only sought independence from Britain to establish freedom for individuals; in addition, they prized equality and justice as the ends of government, and they framed the United States Constitution for that purpose. Wood concluded his analysis by claiming that the promise of civic republicanism, which required individuals to balance self-interest against the interests of the community, was betrayed almost immediately by a rapacious individualism that quickly supplanted virtuous citizens' concern with the common good.
>
> (Kloppenberg 2011: 43)

'As Tocqueville had observed in 1830, and Dewey had confirmed a century later, the secret for arriving at a conception of a shared common good is participation in public life through the process of a lively debate between champions of competing points of view' (ibid.: 32, 41). Excessive individualism returned in the 1960s generation that adopted a more selfish focus on individual rights, a change in attitude that was compatible with the thinking of the right that advocated a greater reliance on market forces and less reliance on the state (Judt 2011: 81–136).

Current state of American democracy

Complete democracy requires not only the participation of all of the electorate in voting but also the existence of a fully-formed public opinion accomplished by effective communications among citizens, experts and politicians that determine the common good. It also requires that politicians are accountable to the citizenry for the policies that they adopt.

The situation in America today is at odds with these prerequisites. Large segments of the American citizenry are apathetic, dislike political participation, and do not vote in congressional and presidential elections. 'The average American doesn't want to be educated; he doesn't want to improve his mind; he doesn't even want to work, consciously at being a good citizen' (Lepore 2012: 55). 'American voters have remained ignorant about political matters despite decades of rising education levels' (Crain 2016: 68). Most citizens do not care about most policies and therefore are content to turn over decision making authority to someone else. A more engaged democracy is simply of no interest to the public (Hibbing and Theiss-Morse 2002).

> In a 1998 survey some 31 percent of respondents agreed with the statement that our government would be run better if decisions were left up to nonelected independent experts rather than politicians or the people. A similar number agreed that government would be run better if decisions were left up to successful business people. Altogether 48% of respondents agreed with one or both of these ideas.
>
> (Sides 2011)

Since the 1960s the dominant ideologies and upper middle-class interests in both the Democratic and Republican parties led to their ignorance of the main economic concerns of the working-class and the lower middle-class voters (Dionne 2004: 13). The apparent response of the poor and the lower middle class was to give up on the government. They have largely disenfranchised themselves. In the 2010 congressional elections, about 30 per cent of the electorate with incomes below the median voted while 68 per cent of the electorate with incomes above the median voted. Consequently, a majority of 60 per cent of voters have incomes above the average (Krugman 2009: 193; Venton 2015: 211).

The results of self-disenfranchisement is a factor that has resulted in governance for the rich. For example, in 2014, two political scientists reported on their multivariate analysis of some 1,779 policy cases over a period between 1981 and 2001 in which a national survey of the general public asked a favour/oppose question and in which there was an income breakdown of the respondents. The researchers correlated the opinions of respondents at the tenth income percentile (quite poor), the fiftieth percentile (median) and the ninetieth percentile (rich). Policy preferences of the ninetieth and fiftieth percentile respondents were analysed with multivariate analysis. Their findings indicated that the majority does not rule – at least not in the causal sense of actually determining policy outcome. When a majority of citizens disagree with economic elites, or with organized interest groups, they generally lose (Gilens and Page 2014: 568, 576).

De facto American oligarchy

An oligarchy is a political regime in which the wealthiest classes are disproportionately powerful and rule for the sake of their private interests (Simpson 2011: 74). The political representatives are drawn largely from the wealthy few who rule over the majority who are poor and not wealthy. The wealthy spend a great deal of money to influence public opinion. The purpose of oligarchic regimes is 'good order' where good order is defined by the temperament of those in power (Cooper 2011: 204). Unlike democracy there is no 'good life' purpose in oligarchic regimes because wealth is an instrumental factor which leaves unanswered the purpose for which it is being amassed (Koivukoski 2011: 183).

Unlike democracy, the administration of justice is unequal owing to limitations on free speech (Cooper 2011: 200). Freedom of speech entails speech that is frank; open; without fear of recrimination; and reveals and uncovers the truth (ibid.: 204). A feature of oligarchy is duplicity to show that the accumulation of wealth by a few is beneficial to the majority (Newell 2011: 9). The limitation on free speech requires a measure of deceit or sophistry or the use of ideology as a cloak for deeper motivations (Bagby 2011: 123).

Oligarchic leaders may not be competent or trained to rule since the acquisition of money has little to do with the knowledge and skills for political leadership and governance. While success in commerce does partake of some substantive claims to excellence of character, these elements of character are not the best virtues for political governance (Newell 2011: 6). They have a conception that blinds them to the fact that the realm of politics is different in kind from the realm of finance (Skultety 2011: 100).

Proposals for reforming American democracy

In this section three proposals are presented for increasing the degree of participation, rendering it more informed and augmenting the practice of democratic deliberation about the common good and discourse on major issues.

Education of citizens for democracy

As noted earlier, a major problem for democracy in the US is voter ignorance of public issues – even among college-educated citizens. For the most part, formal education is liberalized and widely varied and consequently unrelated to democracy and citizen participation in political discourse. This deficiency could be addressed by federal and state programs of education for democracy that would include the following content:

- elements of democracy, its definition, principles, and the nature of the common good;
- language for identifying and addressing rhetoric, specialists' dialects, and propaganda;
- human qualities for thinking and judging alternative policies;
- the psychology of criticism, debate and advocacy;
- ethics and morality of democratic participation;
- principles of economics, political science and philosophy related to the economic system; and
- ecology.

The first five of the foregoing elements were suggested by Canadian philosopher and essayist, John Ralston Saul (Saul 1997: 35, 37, 139, 174). He argued that the existence of a high quality national public education school system for the first dozen or so years of training is the key to democracy where legitimacy lies with the citizen (ibid.: 66). Further, he argued that we now have long periods of free time at the end of our lives which, if transferred to the early stages, could be devoted to a more solid undergraduate humanist education before people entered into a 30- or 35-year career where specialization would narrow their minds (ibid.: 176).

Compulsory voting

As indicated earlier, the large-scale disenfranchisement of the less affluent means that the voters are not representative of the majority of Americans. This problem could be rectified by making voting compulsory, for example, along the lines of the Australian government which levies fines on the electorate who do not vote. Democracy works best when rights are balanced against responsibilities. The right of an equal vote requires a responsibility to vote; without the latter the former is lost. For the vulnerable who could not get out to vote, they would be allowed a mail in vote.

Education for oligarchs

The problem of oligarchy is the excessive inequality of income which gives rise to conflicts between the rich and the poor that cause chaos, extremism and political instability and duplicitous free speech. In this regard, one of Aristotle's remedies was to change the cultural behaviour of oligarchs to moderate their pursuit of wealth by appealing to them as gentlemen to see charity and helping others as honourable and a basis for their self-esteem. To this end, the wealthy were to be educated not to identify their prestige wholly with superior wealth (Newell 2011: 16, 17). In this light the ideal ethics for American oligarchic leaders would include charity and discipline to subordinate personal interests (particularly for money and power) to the interests and values of the of citizens in their constituencies. Among the generic competencies for representatives in Congress, Senate and indeed the Presidency are knowledge of institutional economics, political science and philosophy which are the central disciplines pertaining to the development of policies and management of the economic system for the common good. These competencies and ethics differ substantially from those related to the processes of campaigning for election to office.

Changing the capitalist economic system

Democratic capitalism is about three dynamic systems converging as one: a democratic polity, a capitalist economic system and a moral-cultural system which is pluralistic and in the largest sense liberal (Novak 1991: 14, 57, 58). With democratic capitalism the economic system is governed by the democratic system and its focus is on the common good as determined by the majority of Americans. This is in contrast with present neoliberal capitalism in which the economic system dominates the political system for the purpose of higher growth in per capita GDP. Full employment is prioritized over lower consumer prices that are the hallmark of a consumerist society. Policy details can be found in Venton (2016, 2017), Klein (2014: 114, 115) and Piketty (2014: 309, 513, 515, 516, 544).

American government negotiations with Russia and China

The goal of negotiations is to downsize the military resources and operations of all three empires and to put the remainder of their resources and operation under the auspices of the United Nations as the world's international peace keeping institution. For the US expenditures could be reduced from 3.5 per cent of GDP to 1.5 per cent of GDP. The savings represent a very large peace dividend that could be invested in renewable resources and energy conservation, public education and social welfare for all Americans.

A second goal, like that for America, is to change the capitalist economic systems in Russia and China to reduce their rates of economic growth to rates commensurate with the carrying capacity of the natural resources and ecological systems. As a result of lower economic inequality, there will be some reduction in the production of uneconomic luxury goods and services that have associated high carbon imprints. The overall goal is to move these nations towards economic self-sufficiency. To that end, some of the trade with client states of their empires will be reduced. In other words, competitive free trade will be replaced by collaborative international trade and investment required for balanced economies. Generally, the utilization of technology in these countries, like America, will be oriented to achieving full employment rather than lower prices for consumers. For example, China's needs for full employment of labour may necessitate significantly less productive technology than that employed in America or Russia.

Imbalances between the supply and demand for basic needs in countries can be mitigated to an extent by more liberal collaborative intercountry immigration and trade in natural resources and goods and services. Sovereignty of nation states over their industrial polices, monetary policies, foreign exchange, and tax and fiscal policies will be enhanced. Competitive free trade will be replaced by collaborative free trade to rectify imbalances between the capacity of natural resources and basic consumer needs in each country. Predatory liberalized investment activity will be reduced. These changes will necessitate wholesale changes in the roles and operations of international organizations such as the World Trade Organization (WTO) and the International Monetary Fund (IMF).

Prospects for political revolution in America

The precursor events to political revolution include the Occupy Wall Street movement in 2011 that followed upon the financial crisis of 2008, the 2013

'Blockadia' environmental movement, (Klein 2014: 294–305) as well as reactions to political chaos in the White House in 2017.

Spiritual leadership for change is found in Pope Francis's 'Laudato Si': Praise Be to You' which implicitly contemplates a dialogue about a new economy of democratic capitalism (Venton 2017). Intellectual leadership for progressive change is found in French economist Thomas Piketty's *Capital in the Twenty First Century* and Canadian journalist and activist Naomi Klein's *This Changes Everything: Capitalism vs. the Climate.*

American intellectual leadership for revolutionary progressive change comes from environmental lawyer and former advisor to President Jimmy Carter, James Gustave Speth. His 2012 article 'America the Possible: A Manifesto' lists eleven transformations that are key to building a new political economy that can, if Americans are willing to struggle and sacrifice for it, provide a path to a possible vision by 2050. Speth's vision is that America will have marshalled the economic and political resources to successfully address the challenges to achieve basic social justice, real global security, environmental sustainability, true popular sovereignty and economic democracy. Implementation of prodemocracy reforms will have saved America's politic from corporate control and the power of money. Overall the economy will be governed to ensure broadly-based prosperity and to preserve the integrity and biological richness of the natural world. The US government will be a leader in strengthening the institutions of global governance and international regulation (Speth 2012: 6, 7, 9, 10).

Other American intellectual leaders include economists Joseph Stiglitz, Paul Krugman and political economist Gar Alperovitz who is co-founder of the Democracy Collaborative research institution and a member of the New Economy Movement. Alperovitz has referenced a conference of 1,000 experts who met in Buffalo in 2015 who are working on 'building the theory, the practical experiments, the policies and the politics of a "new economy"' (Alperovitz 2016).

American political leaders include notably Senator Bernie Sanders who is reported to have wide support from millennials and populists who have been savaged in the 'rigged economy'. 'In a poll with a sample size greater than 10,000, three quarters of the American people believe the economy is rigged against them and agreed with the statement "America needs a strong leader to take the country back from the rich and the powerful"' (Sears 2016). The following excerpts from Sanders recent speech on foreign policy to an assembly at Westminster College are instructive.

> Foreign policy is about working with the international community – with Russia, China, India and countries around the world to transform our energy systems away from fossil fuel to energy efficiency and sustainable energy. Foreign policy is not only about military but directly connected to economics; it must take into account the outrageous income and wealth inequality that exists globally and in our own country. So when we talk about foreign

> policy and our belief in democracy, at the very top of our list of concerns is the need to revitalize American democracy to ensure that governmental decisions reflect the interests of a majority of the people. We cannot convincingly promote democracy abroad if we do not live it vigorously at home. A sensible and effective foreign policy recognizes that our safety and welfare is bound up with the safety and welfare of others around the world with 'all the homes and families of all the men and women in all the lands' as Churchill said right here (in his Iron Curtain speech) 70 years ago. One of the most important organizations for promoting a vision of a different world is the United Nations.
>
> (Sanders 2017)

Conclusions

The proposals outlined above for educating Americans and their oligarchic leaders are essential for revitalizing American democracy and reforming its economic system. They are instrumental for realizing the founding fathers' vision of civic republicanism that required individual Americans to balance self-interest against the interests of the community. In the late twentieth century that vision has been supplanted by an economic elite that has dominated the political system. American billionaire Warren Buffet described this dominance in his statement on CNN a few years ago: 'There's been class warfare going on for the last twenty years and my class won … the rich class' (Klein 2017: 81). The proposed legislation for compulsory voting is essential to win the class war for the majority of Americans who are not rich. Such a win is a step towards the partnership of Americans with all the peoples of the world that Bernie Sanders advocates and a subsequent step towards world federalism in the United Nations for international peace and a sustainable environment.

References

Alperowitz, G. (2016) 'Video Address to the No War 2016: Real Security Without Terrorism Conference', 23–26 September, http://worldbeyondwar.org/NoWar2016 (accessed 8 February 2017).

Bagby L. (2011) 'Thucydides and the Importance of Ideology in Conflict', In: Tabachnick D., and Koivukoski T. (eds) *On Oligarchy: Ancient Lessons for Global Politics*, Toronto: University of Toronto Press, pp110–139.

Bradshaw L. (2015) 'Kant, Cosmopolitan Right, and the Prospects for Global Peace', In Koivukoski T. and Tabachnick D. (eds) *The Question of Peace in Modern Political Thought*, Waterloo: Wilfrid Laurier University Press, pp129–148.

Cooper, C. (2011) 'Oligarchy and the Rule of Law', In: Tabachnick D. and Koivukoski T. (eds) *On Oligarchy: Ancient Lessons for Global Politics*, Toronto: University of Toronto Press, pp196–216.

Crain C. (2016) 'None of the Above: The Case against Democracy', *The New Yorker* (7 November) pp67–71.

Dionne E. J. Jr. (2004) *Why Americans Hate Politics*, New York: Simon & Schuster.

Gilens, M. and Page B. (2014) 'Testing Theories of American Politics: Elites, Interest Groups and Average Citizens Perspectives on Politics', *Journal of the American Political Science Association*, vol 12, Issue 13 (Sep) pp564–581.

Hibbing J., and Theiss-Morse E. (2002) *Stealth Democracy: Americans' Beliefs About How Government Should Work*, Cambridge: Cambridge University Press.

Johnson L. (2015) 'Thomas Hobbes on the Path of Peace: Love of Glory versus Realist Foreign Policy', in Koivukoski T. and Tabachnick D. (eds), *The Question of Peace in Modern Political Thought*, Waterloo: Wilfrid Laurier University Press, pp49–70.

Judt T. (2011) *Ill Fares the Land*, New York: Penguin Books.

Klein N. (2014) *This Changes Everything: Capitalism vs. the Climate*, Toronto: Alfred A. Knopf Canada.

Klein N. (2017) *No is Not Enough: Resisting the New Shock Politics and Winning the World We Need*, Toronto: Alfred A. Knopf Canada.

Kloppenberg J. (2011) *Reading Obama: Dreams, Hope, and the American Political Tradition*, Princeton, NJ: Princeton University Press.

Koivukoski T. (2011) 'On Oligarchy: An Ontological Account', in D. Tabachnick and T. Koivukoski (eds), *On Oligarchy: Ancient Lessons for Global Politics*, Toronto: University of Toronto Press, pp178–195.

Krugman P. (2009) *The Conscience of a Liberal*, New York: W. W. Norton and Company.

Lepore J. (2012) 'The Lie Factory: How Politics Became a Business', *The New Yorker* (24 September) pp50–59.

Mann G. (2013) *Disassembly Required: A Field Guide to Actually Existing Capitalism*, Oakland, CA: AK Press.

Newell W. (2011) 'Oligarchy and Oikonomia: Aristotle's Ambivalent Assessment of Private Property', in D. Tabachnick and T. Koivukoski (eds), *On Oligarchy: Ancient Lessons for Global Politics*, Toronto: University of Toronto Press, pp3–23.

Novak M. (1991) *The Spirit of Democratic Capitalism*, New York: Madison Books.

Piketty T. (2014) *Capital in the Twenty-First Century*, Cambridge, MA: Belknap Press of Harvard University.

Sanders B. (2017) 'Remarks at Westminster College', 21 September, www.vox.com/.../2017/.../21/.../bernie-sanders-full-text-transcript-foreign-policy...

Saul J. R. (1997) *The Unconscious Civilization*, New York: The Free Press.

Sears R. (2016) 'Are We on the Verge of a Revolution?' *Toronto Sunday Star* (20 November).

Sides J. (2011) 'Stealth Democracy', *New York Times* (28 September).

Sikkenga J. (2015) 'John Locke's Liberal Path to Peace', in D. Tabachnick and T. Koivukoski (eds), *The Question of Peace in Modern Political Thought*, Waterloo: Wilfrid Laurier University Press, pp71–90.

Simpson P. (2011) 'A Corruption of Oligarchs', in D. Tabachnick and T. Koivukoski (eds), *On Oligarchy: Ancient Lessons for Global Politics*, Toronto: University of Toronto Press, pp70–89.

Skultety S. (2011) 'The Threat of Misguided Elites: Aristotle on Oligarchy', in D. Tabachnick and T. Koivukoski (eds), *Oligarchy: Ancient Lessons for Global Politics*, Toronto: University of Toronto Press, pp90–109.

Speth, J. G. (2012) 'America the Possible: A Manifesto', Part 1: *Orion magazine*, Mar/Apr; Part II: *Orion magazine*, May/June.

Venton P. (2015) 'Radical Changes in Canadian Democracy: For Ecology and the Public Good', in L. Westra, J. Gray and V. Karageorgou (eds), *Ecological Systems Integrity: Governance, Law and Human Rights*. New York: Routledge/Earthscan, pp201–219.

Venton P. (2016) 'Democratic Capitalism for Realizing the Earth Charter Vision', in L. Westra, J. Gray and A. D'Aloia (eds), *The Common Good and Ecological Integrity: Human rights and the support of life*, Routledge/Earthscan, New York, pp192–210.

Venton P. (2017) 'Pope Francis's Ethics for Democratic Capitalism and the Common Good', in L. Westra, J. Gray and F-T. Gottwald (eds), *The Role of Integrity in the Governance of the Commons*, Cham: Springer Nature, pp237–253.

Wikipedia (2013) 'Globalization', https://en.wikipedia.org/wiki/Globalization, accessed 6 May 2013.

Wikipedia (2017a) 'Reform of the United Nations Security Council', https://en.wikipedia.org/wiki/Reform_of_the_United_Nations_Security_Council, accessed 25 January 2017

Wikipedia (2017b) 'Pax Americana', https://en.wikipedia.org/wiki/Pax_Americana, accessed 18 June 2017.

Wood, G. S. (1998) *The Creation of the American Republic 1776–1787*. Durham, NC: University of North Carolina Press.

18 A utopian democratic revolution to overcome flawed democracy and ecological catastrophe

Donato Bergandi

> Utopie, soit. Mais, qu'on ne l'oublie pas, quand elles vont au même but que l'humanité, c'est-à-dire vers le bon, le juste et le vrai, les utopies d'un siècle sont les faits du siècle suivant.
>
> [This may be utopian, but do not forget that utopias, when they have the same aim as humanity – that which is good, just and true – are the facts of the next century.]
>
> (Victor Hugo, *Le Rhin*, 1842)

The political reasons of the ecological catastrophe

What is a revolution? A revolution is a complete and radical change in the order of things; it can be a bloody change or a non-violent cultural, ethical, political, economic total transformation of state of things. Positivism, for example, surely has amounted to a scientific and epistemological revolution where the sciences, part of a progressive project, were the maximum expression of human rationality, able to generate scientific discoveries, the one and only truths. Capitalism too has constituted a social and economic revolution that, by centralizing capital, was able to radically change the division of labour and socioeconomic relationships between men and deeply transformed our relationship with nature. Another cultural revolution can be found in the Conservation of natural resources movement, engendered by Gifford Pinchot (1865–1946), the well-known forester and politician: he fought against the false idea that natural resources were inexhaustible and against their irrational, unproductive waste. He was surely one of the more influential voices affecting the development and conservation of natural resources.

What is a utopia? On the basis of its genealogical history and complexity the word 'utopia' shouldn't be considered as just another word conveying a specific meaning. It should be taken as part of a more complex macro-concept composed of a double morphological and semantic entity: the other part of this macro-concept is constituted by another lexeme, 'dystopia'; hence the macro-concept, 'utopia/dystopia'. In other words, when we talk about the meaning of 'utopia', to better understand it we cannot avoid mentioning its Janusian side, the dystopian side. Why? Because a utopia can easily become or transform into a dystopia.

Or, even more cogently, a real dystopia can be hidden within a seeming utopia. At this point it is legitimate to ask whether, even if some utopian examples exist in literature – Plato (427–347), More (1478–1535), Traiano Boccalini (1556–1613), Bacon (1561–1626), Campanella (1568–1639), Robert Burton (1577–1640), James Harrington (1611–1677) Gabriel de Foigny (1630–1692), Fenelon (1651–1715), Saint Simon (1760–1825), Marx (1818–1883), Aldous Huxley (1894–1963) – are they examples of utopian idealized societies or *intrinsically* dystopian cases? Looking at history, some emblematic cases can help us to better understand, for example, that the utopian Marxian society became a dictatorial partitocratic despotic system, with Joseph Stalin and the communist party in the Soviet Union, or the Nazi racial utopia of the Third Reich (utopia, from the Nazi point of view; dystopia for everyone else) that once it had rid itself of all its enemies would have lasted thousands of years.

Utopia is a protean subject: it leads, with a feeling of adventure and excitement, to uncharted territory, to a boundless forest of possible meanings, open to alternative interpretative variations. From the Greek *οὐ-τόπος*: 'non-place', to *εὖ-τόπος*: 'good-place', we are swung from 'non-existence' to 'ideal', to finish with a 'non-existing ideal place to live'.[1] Semantically and narratively speaking, utopia opens a fictional space for the best (and the worst, dystopic) possible worlds. In fact, the unpredictable complexity of life may harbour unintended consequences or unplanned outcomes with regard to the stated goals to be achieved as well as to the intentional actions. The heterogenesis of ends (Vico [1774] 1948) is consubstantial with the concept of 'utopia' and yields its Janusian character (utopia/dystopia).

Of course, a cultural, ethical, political, economic revolution can assume the character of a utopia or a dystopia. Today we are living the last stages of an ecological dystopia: the last, we hope, for the balance of the ecological systems and the persistence of the evolutionary biosphere's potential, even though some recent, clear indicators tell us that the planet's biodiversity is eroding at a pace unmatched in modern times (Secretariat of the Convention on Biological Diversity 2014; Watson et al. 2016). This erosion concerns animal and plant populations: from invertebrates and micro-organisms, the 'hidden' biodiversity (Wagg et al. 2014), to the vertebrates (Secretariat of the Convention on Biological Diversity 2010; Ceballos, Ehrlich and Dirzo 2017) and insects (Hallmann et al. 2017), the planet's potential biodiversity is shrinking in such a way and pace that we can speak, in no uncertain terms, about 'biological annihilation' due to human activities (Ceballos, Ehrlich and Dirzo 2017).

This environmental dystopia started with a narrative related to a scientific revolution, grounded on reason and logic, a revolution that conveyed a technical and scientific optimism based on belief in the 'progress' of the sciences that was to have engendered a radical transformation of human civilization through, among other things, the systematic use of scientific discoveries by industry (Comte 1832; Geymonat 1971). This positivistic belief in 'The 'magnificent and progressive fate' of the human race' (Leopardi [1836] 2015) has become an economic and technological ecological catastrophe that shares more characters with a dystopia than with an optimistic perfect place where people can live.

So, we can say that in light of this state of affairs a revolution can be utopian when the new situation emerging from the revolutionary dynamics creates relationships between people and between people and the environment that have the propensity to tend towards the 'common good'. The 'common good' in reference to humans is easily understandable as the state of affairs where the interests of all of a community's individuals is pursued: or, in other words, where all the members of the community work together, outweighing partisan concern. Already Aristotle (384–322) looking for what is a good life and what is happiness, arrived at the conclusion that living well and doing well are the characteristic traits of a happy life, a life where the laws aim at the common good or advantage:

> The laws have something to say about everything, their aim being the common interest either of all the citizens, or of the best, or of those in power, or of some other such group. So, in one sense, we call anything just that tends to produce or to preserve happiness and its constituents for the community of a city.
>
> (*Nicomachean Ethics*, V, 1129b–1130a, 82)

This acknowledgement clearly recognizes that a law does not necessarily include the administration of social justice, above all when it preserves some parochial interests to the detriment of the collective good. But what is the common good? This is a very difficult and complex question, and on the answer to this question depends the structure of the political society we work to achieve. Basically, a democratic society, one that pursues global (political, cultural, economical, psychological) 'happiness', the well-being of every citizen, should be able to pursue equally the well-being of the global community.

A dystopian revolution will tend precisely on the contrary to favour, if we follow the political categorization proposed by Jeremy Bentham (1748–1832) and John Stuart Mill (1806–1873), the *sinister interests* of the few that will orient the policies that impact the lives and conduct of the many (Bergandi 2014). Bentham clearly stigmatizes these interests that determine, non-virtuous, 'bad', 'vitious' acts or 'acts of improbity' (Bentham [1815] 1983: §V, 27). Similarly, Mill found that legislation planned for the benefit of the dominant class inevitably is to the detriment of the community as a whole (Mill [1861] 1865: 127–128). In fact, he argued:

> The *positive* evils and dangers of the representative, as of every other form of government, may be reduced to two heads: first, general ignorance and incapacity … secondly, the danger of its being under the influences of interests not identical with the general welfare of the community.
>
> (Mill [1861] 1865: 110; original emphasis)

Those interests that are not coincident with the community's general welfare are considered by Bentham and Mill as misleading, rogue interests that diminish the collective welfare and at the same time augment partisan, class interests, generally the interests of those holding power. For this reason, Mill considers that in a real democracy, to avoid the maximum danger to its existence, we must spurn any type of predominant and exclusive influence and hegemony: 'If we consider as a class, politically speaking, any number of persons who have the same sinister interest – that is, whose direct and apparent interest points towards the same description of bad measures – the desirable object would be that no class, and no combination of classes likely to combine, should be able to exercise a preponderant influence in the government' (Mill [1861] 1865: 128). Especially as this 'predominant influence' can be so hard-hitting as to shape public morality: 'Wherever there is an ascendant class, a large portion of the morality of the country emanates from its class interests, and its feelings of class superiority' (ibid.: 16).

Is it possible that in the same normative foundations of modern democracy lies one of the multiple causes of the current environmental crisis? According to John Locke (1632–1704), one of the sources along with Bentham and Mill of liberalism and modern democracy, people enter into society for the 'preservation of property' (Locke [1689] 1821: 308), and the laws must be designed for no other end but the good of the people (ibid.: 312).[2] Men, naturally free in the state of nature, gave their political power into the hands of society with the 'tacit trust' that this power, grounded on the 'mutual consent' of everyone forming the community, will be used for their good (ibid.: 337). Posing the preservation of property ('goods') as the basis of civil society[3] gave us an axiological meter, the standard to understand what constituted the hierarchical value of Locke's society. This perspective has the merit of having brought about people's freedom to dispose of themselves and their possessions and to contribute to building the culture that produced the Universal Declaration of Human Rights. But, at the same time, the axiological meter of this society to evaluate the non-human world is fundamentally an economic tool for appropriation that, *in essentia*, and in large part, we have inherited. In fact, today this 'property perspective' on the environment, reduced to the unique function of a 'good', still shapes our current worldview (Bergandi 2013). Non-human nature is essentially a means to satisfy human needs, and even if today it is very likely that the narrative shared by many of us about mankind's place in the world is no longer coincident with the details of the Lockean vision, we must recognize that the asymmetrical relationship between people and non-human nature, which seals humanity's ontological and moral primacy, remains substantially unchanged:

> God, who hath given the world to men in common, hath also given them reason to make use of it to the best advantage of life, and convenience. The earth, and all that is therein, is given to men for the support and comfort of their being.
>
> (Locke [1689] 1821: 209)

Our traditional culture still is part of a worldview where people are at the top of evolution and at the center of ecological systems, and everything is at our disposal. From this point of view, our society is not very different or metaphysically far from Locke's seventeenth-century society. On the contrary, our efficacy in appropriating and exploiting the environment is supported by the technological upturn of the twentieth and twenty-first centuries, and this means that if we want to preserve what still remains of the planet's ecological primitive integrity and wilderness, we must make radical, and revolutionary, changes in our perception of humanity's place and role in the world.

A complementary but more radical critique of the elites was made by the Marxists and the anarchists. Marx and Engels (1820–1895) pointed out that the material conditions of life (industry, trade, division of labour) determine the structure of the different social classes, and that the ruling material force of society is also its ruling intellectual and moral force (Marx and Engels [1845–1846] 1976: 59). And, according to Bakunin (1814–1876), always and everywhere, there has been the exploitation of the many by a dominant minority (Bakunin [1882] 1910: 9), and he made a very radical critique of the principle of authority, of the state and policies that 'can turn only to the advantage of a dominant minority of exploiters against the interests of the immense majority in subjection to them' (ibid.: 22).

According to the political theorist Gaetano Mosca (1858–1941), the whole history of mankind, in all societies, at any time, and with very different levels of organization, is the history of a class that rules and a class that is ruled (Mosca 1896: 50). Ruling, organized minorities tend to dominate the unorganized majority (ibid.: 53) and to become hereditary (ibid.: 61) and closed, stationary, crystallized ruling classes (ibid.: 66). If there is the absolute preponderance of a single political force who are the exclusive interpreters of the will of the people, there will be no control to limit 'a natural tendency in those who stand at the head of the social order to abuse their powers' (ibid.: 134).

For the sociologist and economist Vilfredo Pareto (1848–1923), 'The history of man is the history of the continuous replacement of certain elites: as one ascends, another declines' (Pareto 1968: 36); the strongest, energetic new elite supersedes the old one, putting first not its own good, but the good and the rights of the oppressed multitude: a commitment that will be easily and quickly forgotten, or only formally respected, when the new victorious elite take power. Dahl (1915–2014) begins his analysis arguying that 'however wise and worthy the members of a ruling elite entrusted with the power to govern a state may be when they first take power, in a few years or a few generations they are likely to abuse it' (Dahl 1998: 74). So, the 'dark side' of representative government, 'bargaining among political and bureaucratic elites', limits popular participation and control (ibid.: 113).[4] According to the politically engaged American sociologist Charles Wright Mills (1916–1962) a hierarchical, pyramidal power elite – 'those political, economic, and military circles which as an intricate set of overlapping cliques share decisions having at least national consequences' (Mills [1956] 2000: 18) – shares a common psychological and social basis, and is highly coordinated on the basis of its several and convergent interests. Later, Jürgen Habermas puts at the heart of his model of deliberative democracy the influential retroactive effect of the public sphere on political decision-making bodies, but with a big caveat:

it works only 'as long as' the various stakeholders have a deontological posture which is not turned towards their individual success, but towards a common goal of mutual understanding (Habermas 1981: vol. 2, 262).

One of the principal reasons of our feeling of living in a flawed democracy or pseudo-democracy is determined by the fact that the citizens carry out their popular sovereignty by limiting themselves 'to consent', or not, to policies that have been determined elsewhere, by the political parties. In others words, our democracies have become partitocracies.

In many aspects, Athenian direct democracy is very different from our modern democracies, but this historical experience can provide an opportunity to engage in analysis and critique about our actual democracy and its future perspectives, allowing us to find insightful clues from this experience. The officials of the Council (βουλή), with the decision-making roles, were selected by lot, and the offices were attributed by rotation (the officials rotated their responsibilities). The officials served no more than two times in a life, two non-successive years, and they were scrutinized before (to verify their respect for the law) and after (to verify that they have acted properly and were not corrupted) taking office. Any citizen had the right to speak in the Assembly or to propose a law, and the officials were paid equally for their time in service. The decision-making bodies were composed of ordinary citizen, not by professional politicians, and a large part of the population were involved in the political process and in public administration (Farrar 2007: 170–195; Raaflaub 2007: 1–21). The major difference between representative and direct democracy relates to the different political (and psychological) attitudes and posture of the citizen: 'Those who govern can claim legitimacy not because the people have 'consented' to their rule but because every citizen has an opportunity to become a ruler and to make the decisions that affect him and his fellow citizen' (Farrar 2007: 182).

We have learned that the advantages of representative democracy (among others, sovereignty formally lies in the people) also bear a risk and a price to pay, the *sinister interests* of the influential minorities. To face the ecological catastrophe that we are living and a likely democratic disaster, we need a broadening of the population's direct participation in the management of the *res publica*: direct participation structured around the principles of sortition and rotation to prevent the establishment of the *sinister interests* headed by antidemocratic castes. Such principles hinder the will and action of those having particular, *sinister*, rogue interests or having the greatest economic resources. In fact, the selection by lot of officials implies that every citizen has an equal chance to participate as an active subject in making the community's policies. Other salient institutional features of direct democracy include the predominant use of referenda and ballot initiatives, as well as recall procedures if the guidelines for the mandate are not respected. A democracy where all citizens are politically equal.

The *sinister interests* of the elites orient society in a classist, individualistic non-egalitarian way and in a non-environmental direction, so it appears that the political project of empowering the people is one among the few reasonable roads to follow – including to build a direct democratic society that evaluates as a *summum bonum* the integrity of the ecological system as an integral part of human well-being. Is this a utopian democratic revolution? Perhaps. In any case, it is a utopia that is nourished by the incontestable fact that the people, the populations actually 'live' in their territories: they have a 'living sense of place', while in their day-to-day lives they directly suffer with their whole bodies, in their bones, flesh and skins the ecologically devastating effects of the economic preferences of the political-economic caste.

Nor is it clear or evident why our 'representatives' should represent 'our' interests or fight for the preservation of the planetary ecological balance: in the name of what principle? Because they have been elected by the people? Because the common good is at stake? Because we are living in a formal, even if tremendously diluted, democracy? Or, because they are the brave supporters of an Ecological Humanism, environmentally responsible, even though we don't see any warning or precursor signs?

Dystopian conservation of nature

From the political point of view, at the national and international levels, the sustainable development programme supported by governments, UN organizations and, by the way, the majority of transnational companies, is a kind of Trojan horse of the elitist economically oriented interests that are overwhelming the genuine will to protect the integrity of the ecological systems. And in reality, day after day, we observe its worldwide ineffectiveness. Sustainable development is the direct offspring of the conservationism of Gifford Pinchot, where economic considerations were pursued while formally aiming for the 'greatest good to the greatest number for the longest time' (Pinchot 1910: 48). In the long run, this human-centric 'enlightened', utopian vision of the father of the American conservation movement has proven to be the foundational ethico-political ground of dystopian policies that take better care of *human sinister interests* than of the ecological balance of the entire biosphere: human needs have been overlooked compared to the integrity and balance of ecological systems, and the interests of the ruling classes win primacy (Bergandi and Blandin 2012; Bergandi 2014).

The combined, convergent effects of political policies, of an environmentally obscurantist economic model, of an ethically obsolete worldview and of a model of the conservation of nature that ultimately has integrated the pre-eminence of economical imperatives over those of the protection of nature together form an *explosive mixture* that is highly detrimental and catastrophic for the integrity of nature. In fact, the international political policies of an international, oligarchic political-economic caste are grounded in their *sinister interests* that see in the environment nothing but 'a property' that they use as they see fit for their own requirements, that doesn't contemplate the existence of nature *per se* or its careful utilization in respect of natural laws. At the planetary level, a productivist and consumerist economic model is hegemonic and grounded in an obsolete ethical worldview that considers nature as nothing but 'natural resources', which are essentially at the disposition of humanity for its needs and aims. To top it off, a dominant model of the conservation of nature, unable to stop the extinction of species and the ecologically catastrophic fragmentation of the environment by ever more invasive human settlement, is impotent in the face of the ongoing process of nature's destruction.

The International Union for the Conservation of Nature (IUCN), along with the other UN organizations, married into a sustainable development paradigm and has begun to believe, with the good and supportive company of the major transnational corporates, that the preservation of the environment is compatible with growth. 'Protection of Nature is Development' (Bishop et al. 2008; Bergandi 2017). Wow! What a surprise. What an effective ideological mishmash. In *The Future We Want* (UNCSD 2012), of course, 'growth', the most ostracised word of the last thirty years in the UN documents, returns as 'equitable economic growth, creating greater opportunities for all' (ibid.: 4), growth emerging from 'green economy policies in the context of sustainable development and poverty eradication' (ibid.: 56): growth that, with a touch of magic technology, will be able to reduce the emission of greenhouse gases, and with another touch of moral improvement, will be capable of starting responsible industrial production and sustainable patterns of consumption of goods that takes account of the full life-cycle of materials and products. They, among others, 'reaffirm the importance of area-based conservation measures … biodiversity and ecosystem services, are to be conserved through effectively and equitably managed, ecologically representative and well-connected systems of protected areas' (ibid.: 177). But at the same time, ecological functions become 'ecosystem services', fully monetizable and exchangeable on the market. Pinchot's utopian/dystopian programme has become a reality: the development of natural resources and the preservation of waste (Pinchot 1910: 42–46) imply the implementation of an economic model of nature that is assimilated to capital, and must be treated as capital (Hawken et al. 1999).

Having said that, in Pinchot's last book and autobiographical testament, he once again recalls that the fight against monopolies is necessary for the existence of democracy: 'It is the very essence of democracy that the great advantage of each of us is best reached through the common prosperity of all of us. Monopoly is the denial of this great truth' (Pinchot 1947: 506). Pinchot configures his utopia/dystopia on the basis of the democratic control of economic forces and the pursuit of public good. His words allow us to appreciate how much the idealized world of the philosophical father of sustainable development is not only not a reality today, but that it will not happen tomorrow either, unless profound structural changes intervene in our democratic procedures and in our economic model:

> Just as Feudalism, with its tyranny, finally made itself intolerable, so too plutocracy, with its rule over the man by the dollar, with its hardships for the many and its luxury for the few, with its greed and its injustice, must be made to travel the same road. It is time for America and the world to move

> on from a social order in which unregulated profit is the driving force. It is time to move up to a social order in which equality of opportunity will cease to be a dream and actually come to pass. I do not pretend to foretell just what that order will be or by what steps it will be reached. When it comes, I want it to come by development and not by revolution.
>
> When it comes, I hope and believe the new order will be based on co-operation instead of monopoly, on sharing instead of grasping, and that mutual helpfulness will replace the law of the jungle. When it comes, I hope and believe that great unregulated concentrations of wealth, with their enormous power for evil, will no longer be allowed to exist.
>
> (Pinchot 1947: 509)

To consider that the forms of representative democracy that we are experimenting with are the final stage of our societies' political development is tantamount to considering that the political history of humanity is over, and that we can now only make cosmetic changes. Have we really already reached the end of the history of our political institutions? That seems highly questionable. However, if that were indeed the case, in view of the recurring predominance of *sinister interests*, there would be a great deal of concern for the preservation of nature, not yet entirely altered by humankind, which still remains here and there. Only a truly successful democracy that takes into account the 'global interests' of human populations and the balance of the natural systems that support them will be able to cope with the current ecological catastrophe and to preserve the planet's ecological systems and their equilibrium in a practical and sustainable manner.

Notes

1 For an accurate and informative analysis of the theologico-political reasons for the pre-eminent meaning and use of utopias as chimeric, non-existent worlds, see Loty (2006).

2 Notice that for Locke 'property' means the property 'which men have in their persons as well as goods' (Locke [1689] 1821: 340).

3 For Locke, the passage from the state of nature to civil society is determined by a characteristic mix of property (appropriation) and labour: from labour, people gain the right to appropriate what before was in common (Locke [1689] 1821: 218): 'And even amongst us, the hare that any one is hunting, is thought his who pursues her during the chase: for being a beast that is still looked upon as common, and no man's private possession; whoever has employed so much labour about any of that kind, as to find and pursue her, has thereby private possession; whoever has employed so much labour about any of that kind, as to find and pursue her, has thereby removed her from the state of nature, wherein she was common, and hath begun a property.' (ibid.: 212).

4 At the same time, it seems to us that limiting the 'bargaining process' to the political and bureaucratic elites implies an understimation of the role of the economical elites.

References

Aristotle, *Nicomachean Ethics*, translated and edited by R. Crisp, 2004, Cambridge: Cambridge University Press.

Bakunin, M., [1882] 1910, *God and the State*, London: Freedom Press.

Bentham, J., [1815] 1983, 'A Table of the Springs of Action', in *Deontology; Together with A Table of the Springs of Action; and the Article on Utilitarianism*, edited by Amnon Goldworth, Oxford: Clarendon Press.

Bergandi, D., 2013, 'Epilogue: The Epistemic and Practical Circle in an Evolutionary, Ecologically Sustainable Society', 151–158, in Bergandi, D. (ed.), *The Structural Links between Ecology, Evolution and Ethics. The Virtuous Epistemic Circle*, Dordrecht: Springer.

Bergandi, D., 2014, 'Environnement, éthique et politique: les limites d'une démocratie inaboutie et leurs conséquences néfastes sur la protection de la nature', *Éthique publique*, 16(1): 63–81.

Bergandi, D., 2017, 'The Ecological Catastrophe: The Political-Economic Caste as the Origin and Cause of Environmental Destruction and the Pre-Announced Democratic Disaster', 179–189, in L. Westra (eds), *The Role of Integrity in the Governance of the Commons*, Cham: Switzerland, Springer.

Bergandi, D. and Blandin, P., 2012, 'De la protection de la nature au développement durable: genèse d'un oxymore éthique et politique' ['From the Protection of Nature to Sustainable Development: The Genesis of an Ethical and Political Oxymoron'], *Revue d'histoire des sciences*, 65(1): 103–142.

Bishop, J., Kapila, S., Hicks, F., Mitchell, P. and Vorhies, F., 2008, *Building Biodiversity Business*, London: Shell International Limited and the International Union for Conservation of Nature.

Ceballos, G., Ehrlich, P. R., and Dirzo, R., 2017, 'Biological Annihilation via the Ongoing Sixth Mass Extinction Signaled by Vertebrate Population Losses and Declines', *PNAS*, 114(30): E6089–E6096.

Comte, A., 1832, *Cours de philosophie positive*, Paris: Bachelier, Libraire pour les Mathématiques.

Dahl, R. A.,1998, *On Democracy*, New Haven, CT: Yale University Press.

Farrar, C., 2007, 'Power to the People', 170-195, in Raaflaub, K. A., Ober, J., Wallace, R. W., *Origins of Democracy in Ancient Greece*, Berkeley, Los Angeles, London, University of California Press.

Geymonat, L., 1971, 'Il sorgere del positivismo in Francia: Auguste Comte', 427–455, in L. Geymonat, *Storia del pensiero filosofico e scientifico, vol. IV: L'Ottocento*, Milan: Garzanti.

Habermas, J., 1981, *The Theory of Communicative Action*, 2 vols, Boston, MA: Beacon Press.

Hallmann, C. A., 2017, 'More than 75 Percent Decline over 27 Years in Total Flying Insect Biomass in Protected Areas', *PLoS ONE*, 12(10):, e0185809.

Hawken, P., Lovins, A., and Lovins, H., 1999, *Natural Capitalism: Creating the Next Industrial Revolution*, London: Earthscan.

Hugo, V., 1842, *Le Rhin, Lettres à un ami*, vol. 3. Paris: Charpentier, Libraire-Editeur.

Leopardi, G., [1836] 2015, 'Wild Broom: Or, The Flower of the Desert', translated by Steven J. Willett, *Arion*, 23(1): 23–32.

Locke, J., [1689] 1821, *Two Treatises of Government*, London: Whitmore & Fenn.

Loty, L., 2006, 'Which Utopias for Today? Historical Considerations and Propositions for a Dialogical and Paradoxal Alterrealism', *Spaces of Utopia: An Electronic Journal*, I (spring): 19.

Marx, K. and Engels, F., [1845–1846] 1976, *Marx and Engels, Collected works*, vol. 5, London: Lawrence & Wishart.

Mill, J. S., [1861] 1865, *Considerations on Representative Government*, 3rd edn, London: Longman, Green, Longman, Roberts & Green.

Millennium Ecosystem Assessment, 2005, *Ecosystems and Human Well Being: Global Assessment Reports, 1: Current State and Trends*, Washington, DC: Island Press.

Mills, C. W., [1956] 2000, *The Power Elite*, Oxford: Oxford University Press.

Mosca, G., 1896, *The Ruling Class*, New York: McGraw-Hill Book Company.

Pareto, V., 1968, *The Rise and Fall of the Elites*, Totowa, NJ: Bedminster Press.

Pinchot, G., 1910, *The Fight for Conservation*, New York: Doubleday, Page & Company.

Pinchot, G., 1947, *Breaking New Ground*, New York: Harcourt, Brace & Company.

Raaflaub, K. A., 2007, 'Introduction', 1–21, in K. A. Raaflaub, J. Ober and R. W. Wallace, *Origins of Democracy in Ancient Greece*, Berkeley, CA: University of California Press.

Secretariat of the Convention on Biological Diversity, 2010, *Global Biodiversity Outlook 3*, Montréal: Secretariat of the Convention on Biological Diversity.

Secretariat of the Convention on Biological Diversity, 2014, *Global Biodiversity Outlook 4*, Montréal: Secretariat of the Convention on Biological Diversity.

Seth, C., and von Kulessa, R., 2017, *The Idea of Europe: Enlightenment Perspectives*, translated by Seth, C., et al., Cambridge, UK: Open Book Publishers.

UNCSD, 2012, *The Future We Want*, United Nations Conference on Sustainable Development (Rio+20), Rio de Janeiro, Brazil, 20–22 June, New York: UN.

Vico, G., [1774] 1948, *The New Science of Giambattista Vico*, translated by T. G. Bergin and M. H. Fisch, Ithaca, NY: Cornell University Press.

Wagg, C. S., Bender, F., Widmer, F., and van der Heijden, M. G. A., 2014, 'Soil Biodiversity and Soil Community Composition Determine Ecosystem Multifunctionality', *PNAS*, 111(14): 5266–5270.

Watson, J. E. M., Shanahan, D. F., Di Marco, M., Allan, J., Laurance, W. F., Sanderson, E. W., Mackey, B., and Venter, O., 2016, 'Catastrophic Declines in Wilderness Areas Undermine Global Environment Targets', *Current Biology*, 26: 2929–2934.

19 Can the philanthropic imperative enhance international health care?

Paul Carrick

Introduction

In this investigation, I explore the historical, cultural, and future significance of the *philanthropic imperative* for international healthcare funding. The philanthropic imperative states that *we have a duty to reduce sickness and suffering and promote or restore, as generously as possible, the health of our fellow human beings*. In addition, I aim to discover the roots of this principle for a deeper understanding of the conceptual frameworks that have moved individuals, communities, and nation-states to fund health care from ancient to modern times.

Elsewhere I have established that this preeminent humanitarian principle is implicitly at work in fostering health care funding on a national level (Carrick 2007: 38–40). Now I expand my inquiry to the broader international level. So, I ask: if the Philanthropic Imperative were recognized and openly endorsed by the international community of nations, what would happen? If these nations continue to work together to achieve humane medical care through cooperative arrangements with each other, and with such groups as the World Health Organization, the International Red Cross, or Doctors Without Borders, would overall health care funding be enhanced by the global endorsement of the philanthropic imperative?

I argue that the answer to this question is a qualified 'yes'. That is, the international endorsement of the philanthropic imperative – which I take to be a sufficient condition at the heart of any humane program of publicly funded medical care – would go a long way toward strengthening and sustaining sound programs of global public health well into the twenty-first century.[1]

What is the philanthropic imperative?

Let us take a closer look at four powerful action-guiding motives that have supported the private or public funding of health care in western Europe since at least the beginning of the Industrial Revolution around 1750. By way of illustration, the *redemptive motive* historically encouraged the funding of hospitals (or, in earlier times, hospices or related care-giving efforts) by those individuals or groups which aimed to win divine redemption for their souls, or the favour of heaven for their communities through the providence of God. Of these four motives, the redemptive is the only one which is manifestly religious in nature. The other three are ostensibly secular; yet they may also carry a religious connotation depending on context.

Second, the *utilitarian motive* is illustrated by the funding of strategically erected medical infirmaries for sick or wounded soldiers. The swift and efficient return of injured warriors to the battlefield, or at very least their rapid replacement by fresh recruits, would help maximize and protect – all things being equal – the common good of any nation facing an aggressive enemy bent on war.[2]

In contrast, the *prudential motive* is somewhat narrower in scope. It aims to adopt an appropriate means by which to wisely and efficiently accomplish a specific end. For example, the prudential motive is illustrated in the funding and deployment of preventive medicine measures such as quarantines. Again, this involves the application of appropriate means to resist or halt a deadly epidemic terrorizing the land.

Finally, there is the *charitable motive*. For example, this motive has animated the funding of public health measures such as inoculations. Inoculations are but one example of providing medical goods or services often given freely to the masses by individuals or institutions experiencing a deeply felt impulse to render a helping hand to the sick and suffering. Or to succour, heal, or comfort those at risk of contracting the onslaught of a communicable or non-communicable disease. Hence, the charitable motive could also involve sponsoring public health warnings or general education campaigns designed to fight such scourges as syphilis, contaminated food, unsanitary drinking water, or ringworm.

In sum, uniting our quartet of acting-guiding motives is the humanitarian vision that good will toward one's fellow man involves consideration of a higher-order imperative, an imperative that draws upon some of our most deeply felt religious or secular sympathies – namely, the philanthropic imperative. Again, it asserts that *we have a duty to reduce sickness and suffering and promote or restore, as generously as possible, the health of our fellow human beings*. Logically considered, I find that any one of these four motives constitute a sufficient condition to render some form of mutual aid. That is, if present, each one singly or jointly functions psychologically and morally to promote various types of individual, community, or institutional healing aimed at the relief of human suffering (Carrick 2012: 56).

Furthermore, I contend that the philanthropic imperative embodies a unifying Hippocratic synthesis. To better appreciate this fact, recall the prescient words of the Hippocratic author of *Precepts* who, in 350 bce Greece, declared:

> And if there be an opportunity to serving one who is a stranger in financial straits, give full assistance to all such. For where there is love of man [*philanthropia*], there is also love of the art.
>
> (Carrick 2007: 38)[3]

Nevertheless, it is fair to ask: what sort of meta-ethical theory supports and grounds the Philanthropic Imperative? For example, does it belong to the results-based utilitarian camp, the duty-based deontological camp, or the virtue-based eudaimonistic camp? My answer is that this eclectic imperative belongs to all three. The philanthropic imperative involves the commitment of *beneficence* entailing positive welfare duties to do good in the world associated with the utilitarian perspective. Also, it involves the duty of *respect for persons* entailing autonomy (self-determination) associated with the deontological perspective. Lastly, it involves the virtue of *justice* (giving to each his due) associated with the positive character trait of integrity. Hence, 'integrity', as defined here, is a harmonizing, higher-order virtue encouraging truth-telling, promise-keeping, and fairness in the conservation and distribution of the Earth's vital resources among the diverse peoples and nations of the world.

The philanthropic imperative: promises and challenges

But how practicable is the philanthropic imperative in solving real world problems? This question will be addressed shortly. First, are there specific philosophers or theologians who would find the philanthropic imperative acceptable? Indeed so. Among several that I could cite, Immanuel Kant, Thich Nhat Hanh and Peter Singer come readily to mind.

Immanuel Kant

As for Immanuel Kant, did not this intellectual giant declare that acts of charity[4] are only 'imperfect duties'? This, since such acts cannot strictly speaking be required by his categorical imperative as acts performed from duty?[5] Quite so. Yet Kant also allows that acts of charity are frequently morally praiseworthy. For if willed with pure intentions, charitable deeds treat others as 'ends-in-themselves'.[6] That is, as persons valuable in and of themselves and not as mere tools to our private ends (Kant [1785] 1976: 47–48). Thus, we see that the charitable motive associated with the philanthropic imperative would also be acceptable to Kant's philosophy provided one's motives are neither self-serving nor wholly results-based. Indeed, Kant would require that *respect for persons* (e.g. healing others for their own sake and not for ulterior motives) must be the moral caregiver's primary motivation. Contrary to a common misinterpretation, then, Kant does not reject acts of beneficence but rather encourages them so long as these acts are determined by the moral law (i.e. his categorical imperative).

Thich Nhat Hanh

As for Thich Nhat Hanh, does not this Zen Buddhist monk teach that just as we live, breath, and act intentionally in the world, we can also be undone by what the Buddha called the three poisons: anger, greed, and ignorance?

Specifically, *ignorance* of the pain and suffering of others (not merely ourselves) is something we need to wake-up to. In addition, *greed* or selfishness can make us think we need more food, shelter, medical supplies, money – than we actually do. Lastly, *anger* at ourselves – because we are watering the unquenchable 'seeds'[7] of avarice, materialism, and hate within our bodies, instead of nurturing the life-affirming seeds of lovingkindness, compassion, and openness toward others – causes us to be alienated from both self and other. Hence, this Vietnamese guru declares: 'The "right action" is the action that results in the fires of hatred and violence being extinguished' (Hanh 2005: 108).

So, it is fair to ask: what do the psychological dynamics of hatred, violence, alienation, anger, greed, and ignorance have to do with the validity of the philanthropic imperative? The answer is not far to seek. By acting with heartfelt charity, lovingkindness and compassion toward others, as the philanthropic imperative ideally requires, we must first put our own spiritual houses in order. That is, we cannot meaningfully attend at the national or international level to the medical and related needs of others – giving away to others valuable resources we could keep for ourselves – until we confront two profound questions invited by Thich Nhat Hanh's philosophy of non-violence:

First, are my philanthropic acts or intentions free of hate, suspicion, fear, or prejudice toward the stranger who is living and suffering in faraway lands, lands that I may never personally visit, inhabit, or explore?

Second, if I fail to provide medicine, food, shelter and related necessities to the hurting, suffering poor, often scratching out a bare subsistence living in the developing South, does my failure to act constitute or symbolize – by omission – a reflexive, indirect act of violence, carelessness, or disrespect toward 'the least of these'?[8]

In sum, meditating on the implications of these two questions constitutes an essential preliminary step to a personal or institutional commitment to helping fund international health care programs as generously as possible. Taken together, they provide a vital exercise in promoting mindfulness and moral clarity for anyone aiming to approach the demands of the philanthropic imperative with a clear conscience, strong mind, and open heart.

Peter Singer

What about Peter Singer? Does not this Australian utilitarian philosopher and animal rights advocate espouse a personal and universal commitment to international charitable giving? This, to benefit the sick and suffering – especially those experiencing extreme poverty? Indeed, Singer embraces the following altruistic principle: 'If it is in our power to prevent something bad from happening without thereby sacrificing anything of comparable moral importance, we ought, morally, to do it' (Singer 1972: 230).[9]

Specifically, Singer suggests that each citizen who is financially able ought to donate at least 1 per cent of his or her annual income to international healthcare projects (Singer 2016: 159). This is his bare minimum of suggested giving – assuming one lives above the US poverty line (Singer adopts the World Bank's definition of 'extreme poverty' as living on less than $1.90 per day).

Furthermore, Singer's charity organization, called 'The Life You Can Save', recommends and monitors approximately 20 specific charities that he urges everyone above the poverty line to consider supporting. In fact, for wealthy individuals, Singer advocates a much more generous scale of annual giving than merely 1 per cent. He advocates a sliding scale of philanthropy based on one's annual income that can top 10 per cent (for details, see www.petersinger.info/faq).

In support of his philosophy, Singer discloses that he and his wife currently give one-third of their annual income to charity; he states that they would like one day to give one-half (see www.thelifeyoucansave.com). Singer observes: 'At the UN Millennium Summit [in 2001], the nations of the world committed themselves to a set of targets, prominent among which was halving the number of people living in poverty by 2015' (Singer 2016: 160). Thanks in part to the vigilant monitoring and vetting methods of 'effective altruism', the upshot is that the current number of people dwelling in extreme poverty worldwide has dropped from over 1 billion people in 2001 to about 702 million in 2016. Most of these impoverished individuals are living in Africa and Southern Asia. Over one-third are children (ibid.: 162, 174).

Hence, with thousands of children dying annually from poverty-related causes, there is obviously a pressing need internationally for safe food and secure shelter, clean drinking water and reliable sanitation, and immediate medical care against infectious diseases. So, looking to the philanthropic imperative, how much money should individuals, community organizations, or nation-states give?

For individuals, I contend that private citizens who are able should give at least 1 per cent of their annual incomes to either national and international organizations such as those on Singer's preferred list. Such groups as the Salvation Army, International Red Cross, Doctors Without Borders, or World Health Organization would certainly qualify.

Moreover, almost all wealthier Northern nations ought to revise their annual budgets to provide significantly more international development, non-military foreign aid than they currently do. Specifically, they should give vetted international charities at least 2 per cent of their annual budgets. Yet few wealthier Northern nations are meeting this increased level of giving in 2016. Therefore, despite its fund-raising potential, the moral force and potential positive impact of the philanthropic imperative has so far yet to be realized. Still, I contend that were this principle openly discussed and embraced as a guiding ethical imperative by the world community of nations (which, obviously, has not yet happened), then the philanthropic imperative would gradually animate a positive increase in funding levels for international public health care programmes of every stripe.[10]

The Earth Charter and sceptical challenges

In weighing the pertinence and fund-raising potential of the philanthropic imperative, it is instructive to recall that several key sections of the Earth Charter stand in overall agreement with its aims. But first, what is the Earth Charter?

According to Laura Westra:

> The Earth Charter, as a cosmopolitan moral perspective, includes respect for the preconditions of life – a recognition of the interconnectedness of all life. [At bottom] ... it emphasizes a respect for communities and peoples that is basic to global ethics. It is a declaration of fundamental principles for creating a just, sustainable, and peaceful society in the twenty-first century.
>
> (Westra 2012: 608)

Specifically, three passages of the Earth Charter speak to our mutual responsibility for promoting the health and welfare of all people, thereby tacitly intersecting with the basic aims of the philanthropic imperative.

First, under 'Preamble', the section on universal responsibility states:

> To realize these aspirations, we must decide to live with a sense of universal responsibility, identifying ourselves with the whole Earth community as well as our local communities. We are at once citizens of different nations and of one world in which the local and global are linked. *Everyone shares responsibility for the present and future well-being of the human family and larger living world.*
>
> (Earth Charter 2000: 1, emphasis added)

Second, under 'Ecological Integrity', article 7e states: '*Ensure universal access to health care* that fosters reproductive health and responsible reproduction' (Earth Charter 2000: 3, emphasis added).

Third, under 'Social and Economic Justice', article 12 states: 'Uphold the right of all, without discrimination, to a natural and social environment *supportive of human dignity, bodily health, and spiritual well-being* with special attention to the rights of indigenous peoples and minorities' (Earth Charter 2000: 4, emphasis added). Hence, the mutual affinity of the philanthropic imperative and the Earth Charter cannot be ignored: together they invite opportunities for the enhanced funding of international public health care programs both at home and abroad.

But there is a serious sticking point. What about the problem of local government corruption? Sceptics scorn the combined efforts of international philanthropy (i.e. 'foreign aid') as naïve, wasteful, and misdirected. This, because corrupt local governments sometimes steal these charitable contributions,[11] depriving their own sick and suffering citizens of the medical care they so desperately need. Nevertheless, I concur with Thomas Pogge's sobering rejoinder:

> The populations of most of the countries in which severe poverty persists or increases do not 'govern themselves' poorly, but *are* very poorly governed, and much against their will. They are helplessly exposed to such 'government' because the rich states recognize their rulers as entitled to rule on the

> basis of effective power alone. We pay these rulers for their people's resources ... and we eagerly sell them advanced weaponry on which their continued rule all too often depends.
>
> (Pogge [2007] 2013: 540)

At bottom, when wealthy nations attempt to 'explain away' the misery of chronically poor nations, such as Haiti or Liberia, as merely the result of local corruption – thereby excusing themselves from any responsibility or any increase in philanthropy – this frequently overlooks their own institutional co-dependency. If so, the enabling behaviour of the wealthier nations must be exposed and corrected in order to fully restore trust and integrity to the philanthropic process. In addition, we are reminded that human suffering associated with malnutrition and extreme poverty is not simply due to a lack of available food and water (Easterbrook 2009: 12). Rather, the larger difficulties we face are: (a) enforcing effective altruism; (b) the timely and unobstructed distribution of life-sustaining resources; and (c) appropriate technology transfer (including education) to hurting men, women, and children wherever we encounter them around the globe. This is undeniably a tall order, one sure to test our most earnest and well-intended philanthropic efforts.

Neo-Malthusian objections and replies

Any serious discussion of the merits of enhancing or increasing medical or food aid to distant people facing extreme poverty must confront the added, persistent objections of Neo-Malthusian critics such as the biologist and utilitarian Garrett Hardin (Hardin [1974] 2012: 296–305). Hardin advances a lifeboat metaphor to champion at least two controversial objections to food or medical aid. The first I call his objection from futility. The second I call his objection from scarcity. Let's take them each in turn.

The *objection from futility* asserts that a thoroughgoing utilitarian analysis of the aftermath of food or medical aid to poor, overpopulated nations will show – contrary to what many good Samaritans may think – that in the long run it is morally justifiable to refuse such humanitarian aid to these drowning, impoverished people. Why? Because in consequence of receiving such aid, once they recover their health and strength, they will again procreate – producing even more children. But producing even more children will tragically restart the vicious cycle of malnutrition and death all over again. For these recovering overpopulated peoples cannot survive beyond the carrying-capacity of their local environment. Tragically, as this vicious cycle of depletion and replenishment resumes, it causes an even greater balance of misery over happiness than if no food or medical aid had been rendered in first place. In short, seemingly praiseworthy efforts to rescue these drowning unfortunates actually make things worse, not better (see also Hardin [1986] 2012: 272–282). Again, upon recovering their health and strength, these unfortunate people overpopulate and overuse their limited land, thereby unwittingly ushering in more starvation and more death.

Relatedly, the *objection from scarcity* asserts that wealthier Northern nations set themselves up for catastrophe if they give away their scarce reserves of food or medical supplies to the struggling poor. For don't wealthier nations need a secure stockpile of reserves for themselves? Don't they want to continue to flourish in the often hostile, unpredictable, Hobbesian world of finite resources? If so, by giving away their vital medical supplies, wealthier nations will almost certainly put themselves at greater risk of perishing. After all, they daily face the miserable poor swimming around their secure well-supplied lifeboats begging or threatening to climb aboard. Tragically, lending a helping hand or allowing the desperate poor to climb aboard could very well sink the Northern nations' own chances of survival.

These two objections are valid only if Hardin's 'lifeboat metaphor', and related 'environmental commons' analogy, are factually sound. But they are not sound. Indeed, Hardin's analysis commits the twin fallacies of oversimplification and hasty conclusion.

What is *oversimplified* becomes obvious when recalling that his lifeboat metaphor implies that wealthy Northern nations are existing as if they inhabit big isolated flourishing lifeboats. These comparatively secure lifeboats are separated by churning seas from the poorer floundering nations of the South. In contrast, inhabitants of poor nations are pictured as barely afloat, drowning before the very eyes of the wealthy. Yet wealthy and poor alike are tossed about in these vast oceans, the so-called 'commons'. The commons represents finite resources of the planet (e.g. oil, pastures, fresh water) –resources that are diminishing in unpredictable ways due to overuse and overpopulation.

Yet Hardin and his followers studiously ignore that individual nation-states today are rarely separated and isolated (Murdoch and Oaten [1975] 2012: 307–308). That is, nations trade goods and services, communicate regularly, and harvest vital planetary resources together. Also, Hardin insists if we let the desperate poor climb aboard and partake of our food or medical resources, they will likely swamp our well-stocked, secure lifeboat. Therefore, if Northern nations act charitably, eventually rich and poor alike will drown by overloading and overusing vital resources. Again, nation-states North and South are interconnected as never before in history in trade, communication, commerce, security agreements, etc. Many of these agreements or treatises are backed by international law. Hence, Hardin's analysis is over-simplified and uniformed. His lifeboat metaphor and related commons analogy is both doubtful and unsound.

In addition, Hardin's *hasty conclusion* just here is also exposed. His heartless, non-interventionist argument rests on the assumption that he somehow knows in advance how many essential life-sustaining resources a wealthy nation needs to stockpile in order to ensure its own survival against future environmental catastrophes. But Hardin cannot know in this detail what awaits nation-states in their future environmental struggles. So, Hardin's claim of doomsday consequences if wealthier nations extend medical or food-aid to struggling poorer peoples is overdetermined. At bottom, Hardin not only over-presses his evidence in favour of heartless non-intervention. He also implicitly commits the related *fallacy of unknowable statistic*: he cannot really know in any detail what percentage of food and medicine a nation must stockpile to avert future environmental catastrophes.

The philanthropic imperative at work

Given the lingering criticisms of Hardin and others who argue that programs of international foreign aid are largely wasteful, futile, and even dangerous – due to a combination of human overpopulation, episodic intergovernmental corruption, vital resource depletion, and anthropogenic planetary degradation – how can we in good conscience support increasing global levels of philanthropy required by the philanthropic imperative?

One possible answer comes from Microsoft billionaire Bill Gates. Along with his wife Melinda, they donate annually over one-half of their earnings to charities around the world through the Gates Foundation. In a brief article, Gates declares:

> Foreign aid is often in the hot seat, but today the heat is cranked up especially high. The U.S. government, one of the world's most influential donors, is considering dramatic cuts to health and development programs around the world. I understand why some Americans watch their tax dollars going overseas and wonder why we're not spending them at home. Here is my answer. These projects keep Americans safe. And by promoting health, security, and economic opportunity, they stabilize vulnerable parts of the world … [Foreign aid is] protecting Americans, saving lives: aid delivers phenomenal benefits, and for a bargain. It represents less than 1 percent of the federal budget, not even a penny out of every dollar.
>
> (Gates 2017: 39)

In the main, I think Gates is right to stress the importance of American foreign aid and related efforts which he fears may soon be decreased. Indeed, based on this investigation, the US should consider doubling its foreign aid contribution to at least 2 per cent. This suggested increase would be justified by that part of the philanthropic imperative which requires us to give 'as generously as possible'. Furthermore, I contend that the United States and other comparatively wealthy Northern countries should each increase their international public health care funding without necessarily looking for a *quid pro quo*. Gates declares that we can usually count on some positive return from our foreign aid investment. Again, he may be right. Yet we need to frankly acknowledge that there are no guarantees.

So, I ask: isn't there an even higher rationale, a higher *raison d'être* for aiding one's fellow human beings in dire straits? Especially the sick and the suffering trying to scratch out a living in impoverished environmental circumstances around the globe? Contrary to Gates's narrowly pragmatic, bean-counting rationale, isn't there something more sublime that wins our moral assent to the philanthropic imperative?

Indeed, I think there is. Immanuel Kant, Thich Nhat Hanh, and Peter Singer all remind us. The Earth Charter beckons us. And, indeed, as I have suggested here: charity is truly its own reward. Especially when the funding of public health care –nationally or internationally – is given freely, for the right reasons, with a calm mind and compassionate heart. Expecting nothing back in return. But working for positive change through direct action, mutual aid, and a commitment to the philanthropic imperative.

Notes

1 My conjecture is both normative and empirical. It is *normative* in that I define and defend the philanthropic imperative as a sound principle of ethics. It is *empirical* in that I hypothesize that if the philanthropic imperative were endorsed internationally, it would enhance and encourage global funding of public health care programmes worldwide.

2 The term 'utilitarian' is being used in the pre-philosophical sense: an act has utility if it is thought to promote the common good. I am not invoking the classical utilitarian theories of Jeremy Bentham or J. S. Mill.

3 What do I mean by 'philanthropy'? Philanthropy literally means 'loving mankind' (from *phil* = love, and *anthropos* = mankind). *Webster's Third International Dictionary* defines it as: (1) 'good will toward one's fellow man, especially as expressed through active efforts to promote human welfare;' or (2) 'an act … of generosity; a contribution made in the spirit of humanitarianism'. See also Carrick (2001: 104–105; 2007: 38–40).

4 By 'charity', I mean acts involving giving away to the other (including the stranger) things of real value (e.g. medical resources, food, money, tools, time, expertise, etc.) that one could otherwise keep for one's self.

5 An example of a 'perfect duty' for Kant is paying back a personal loan one has promised to repay. To stiff the lender is behaviour that could not be universalized as a maxim without logical contradiction. A non-consequentialist, ethical absolutist, and formalist, Kant would find any such cheating blameworthy: it fails to conform to his categorical imperative. See Kant ([1785] 1976: 18–19).

6 See Kant's second form of his categorical imperative, which (paraphrased) states: Treat all persons as ends-in-themselves, never as mere means (ibid.: 46).

7 Hanh uses the colourful metaphor of life-affirming and life-destroying 'seeds' (i.e. desires, volitions, choices) that enter into one's consciousness. These become 'seeds of conduct' only if we act on them. See Hanh (2005: 60–61).

8 See Matthew 25:40. Matthew warns that God is especially on the side of the hurting and downtrodden, not the comfortable, self-satisfied elite. See also: Leviticus 19:18.

9 Singer's impartial, altruistic principle has generated a rich swirl of debate. He confronts his critics at Singer (2002: 153–172).

10 As of 2016, the International Committee of the Red Cross is sustained by approximately 1.5 billion in Swiss francs. The bulk of its field budget is currently expended in Africa (Ebola, malaria, HIV) and the Middle East (civil wars Syria, Iraq). Its revenue sources include governments (85.1%), the European Commission (9.7%), and public and private sources (3.2%). Operated by over 80 countries, it involves 17 million volunteers while upholding the humanitarian laws of the Geneva Conventions (1949). See www.icrc.org. See, too, the World Health Organization's website at www.who.org. Its 17 'sustainable development goals' include climate change (goal 13), equitable access to health care (10), and clean water and sanitation (6).

11 Pogge answers his many critics at Pogge ([2007] 2013: 536).

References

Carrick, P. (2001) *Medical Ethics in the Ancient World*, Georgetown University Press, Washington, DC.
Carrick, P. (2007) 'The Public Funding of Health Care: A Brief Historical Overview of Principles, Practices, and Motives', in N. Maclean (ed.), *Distributing Health Care: Principles, Practices, and Policies*, Academic Imprint, Exeter, pp. 11–40.
Carrick, P. (2012) 'Aldo Leopold's Concept of Land Health: Implications for Sound Public Health Policy', in C. Soskolne, D. Spady, L. Westra (eds), *Human Health and Ecological Integrity: Ethics, Law, and Human Rights*, Routledge, New York, pp. 56–65.
Earth Charter (2000) 'The Earth Charter', www.earthcharter.org (accessed 16 November 2004), pp. 1–6.
Easterbrook, G. (2009) 'Norman Borlaug: The Man Who Defused the "Population Bomb"', *Wall Street Journal*, 16 September, p. 12.
Gates, B. (2017) 'How to Keep America Safe', *Time Magazine*, 15 May, p. 39.
Hanh, T. N. (2005) *Calming the Fearful Mind: A Zen Response to Terrorism*, Parallax Press, Berkeley, CA.
Hardin, G. ([1974] 2012) 'Lifeboat Ethics', in L. Pojman and P. Pojman (eds), *Environmental Ethics: Readings in Theory and Application*, sixth edition, Wadsworth, Belmont, CA, pp. 296–305.
Hardin, G. ([1986] 2012) 'The Tragedy of the Commons', in L. Pojman and P. Pojman (eds), *Environmental Ethics: Readings in Theory and Application*, sixth edition, Wadsworth, Belmont, CA, pp. 272–281.
Kant, I. ([1785] 1976) *Foundations of the Metaphysics of Morals*, Beck, L. W. (trans), Bobbs-Merrill, Indianapolis, IN.
Murdoch W. and Oaten, A. ([1975] 2012) 'Population and Food: A Critique of Lifeboat Ethics', in L. Pojman and P. Pojman (eds), *Environmental Ethics: Readings in Theory and Application*, sixth edition, Wadsworth, Belmont, CA, pp. 306–311.
Pogge, T. ([2007] 2013) 'World Poverty and Human Rights', in J. Boss (ed.), *Analyzing Moral Issues*, sixth edition, New York, pp. 534–540.
Singer, P. (1972) 'Famine, Affluence, and Morality', *Philosophy and Public Affairs*, vol 1, no 3, pp. 229–243.
Singer, P. (2002) *One World: The Ethics of Globalization*, Yale University Press, New Haven, CT.
Singer, P. (2016) *Ethics in the Real World: 82 Brief Essays on Things That Matter*, Princeton University Press, Princeton, NJ.
Westra, L. (2012) 'The Earth Charter: From Global Ethics to International Law Instrument', in L. Pojman and P. Pojman (eds), *Environmental Ethics: Readings in Theory and Application*, sixth edition, Wadsworth, Belmont, CA, pp. 606–608.

20 The uses of poetry to effect positive climate-change policy

Joan Gibb Engel

I.

I am making a claim for the importance of poetry in dealing with climate crisis and I'm up against some widespread prejudice. In April 2017, the *New York Times* published a lengthy article on the increased importance of protest poetry since the election of Donald Trump (Alter 2017: A1, A11). The print version of this article, while elaborating on poetry of racism, civic identity, and social justice, did not even mention the poetry of climate change. The late literary critic John Felstiner, author of *Can Poetry Save the Earth*, defined the problem. 'Realistically', he asks in that book, 'what can poetry say, much less do, about global warming, seas rising, species endangered, water and air polluted, wilderness road-ridden, rainforests razed, along with strip mining and mountaintop removal, clearcutting, overfishing, overeating, overconsumption, overdevelopment, overpopulation, and so on and on?' And he answers, 'Well, next to nothing. "Poetry" and "policy" make an awkward half-rhyme at best'. He does add, 'Yet next to nothing would still be something' (Felsteiner 2009: 7)

Recalling the verses typically memorized in grade school, you may feel that poetry is far too flimsy for the serious battles we face when dealing with climatic and other environmental matters. I fear that the familiar 'Trees' by Joyce Kilmer continues to represent our idea of the position of poets as they wrestle with human/nature relationships. Composed before World War I, with the horror of holocausts and the threat of nuclear bombs yet to come, it separated trees, God, and humans – a beneficent God, a pious nature, and a worshipful politically detached humanity represented by the poet/scribe (Kilmer 1913: 160). This stance is difficult to relate to today.

Yet few would disagree, I believe, with my saying that poetry remains a storehouse for the things of nature that we value and are losing or have lost. It is through poetry that we express the meaningful, humane understandings that our species has constructed from its contact with the things of Earth. A favourite poem of mine, 'Advice to a Prophet' by Richard Wilbur, written in 1966 to dissuade the world from the use of nuclear weapons, argues that, should we lose these things of nature, we would also lose our humanity; our love would leave with the absent rose, our 'lofty' and 'long-standing' with the absent trees (Wilbur 2004).

Because of this threat, because you and I, we fortunate ones, have had love and lofty, our obligation is to preserve what of nature remains lest we lose the interior as well as the exterior world. This involves poetry. For it is in the nature of poetry to remember, to layer the present with the past as Wordsworth does in 'Tintern Abbey'. It is in the nature of poetry to mourn the passing of time as in the famous line 'where are the snows of yesteryear'. And so it becomes natural and useful to turn to poetry to mourn the multitude of losses small and large occasioned by climate change, the everyday taken-for-granted 'heres' that will no longer exist, as poet Christian McEwen does in 'In Memoriam', citing, among other losses, a neighbouring gardener forced indoors, a flower once casually picked, waiting in rain for the school bus (McEwen 2008: 17).

Already we who value nature-related experiences have difficulty talking to persons for whom these 'ordinary' pleasures are remote or missing, who dwell mentally in a technology-driven world where garden flowers, leaves, grass, animals are but dots on a screen. Writers who deal with climate change and other threats to the environment can no longer assume a common appreciation of the inherent glories of nature. The poet Mary Oliver addresses this problem in 'From This River, When I Was a Child, I Used to Drink' (Oliver 2008: 44) The speaker in her poem comes back to a beloved river that is dying and is asked to explain how she feels. When she says she will grieve for future generations who will not know the nature of a river, how it can be 'a friend, a companion, a hint of heaven' she is accused of 'overplaying' the situation. Oliver gives no ground. She repeats her statement without change.

This same problem is dealt with in a different manner by W. S. Merwin in 'Witness', a poem frequently anthologized for its relevance (Merwin 1988). To tell 'what the forests were like' Merwin claims to have to speak 'in a forgotten language'. This can be interpreted in different ways. I see it as saying that language remains essential but its forms have to adjust to the realization that the magnitude of the world's change has also changed its listeners. If humans have lost the connections between words and phenomena, new words, new forms, must be invented.

II.

The danger of rising carbon-dioxide levels is a difficult matter to convey in any form. Unlike its consequences – flooding, drought, species extinction, dead coral, storms, refugees – these tragedies are easy to describe – climate change arrives on little cat feet like Sandburg's fog. Number, the truth-medium of science, fails to impart the seriousness of what is occurring. A 1.3 degree Fahrenheit increase in temperature since 1895? How bad can that be?

Yet, far from being flimsy, poetry packs a punch missing in so many of our statistic-heavy presentations. Admittedly, we need graphs and equations and all the other instruments in a scientist's toolbox, yet a two-sentence poem by Jane Hirshfield warns of an unacknowledged climate-change disaster better than all those stairstep lines. In 'Global Warming' she quotes the explorer Captain James

Cook, who wrote that the Australian natives paid no attention to the arrival of his ship, 'Unable, it seems, to fear what was too large to be comprehended' (Hirshfield 2013: 129).

Although we often think of poetry as a passive art, it is really no stranger to environmental protest. For poetry has a way of revealing truth in the midst of lies. When nature's 'heres' began to disappear big time, such as during the enclosure movement of the 1800's in Great Britain, poets rose to object. John Clare in 'The Mores' melded the plight of dispossessed peasants and nature's ruin:

> Each little tyrant with his little sign
> Shows where man claims, earth glows no more divine
>
> (Clare 2005)

The oft-quoted lines:

> The law locks up the man or woman
> Who steals the goose from off the common
> But leaves the greater villain loose
> Who steals the common from the goose

are from a handbill distributed to prevent support for the enclosure movement (Boyle 2003: 33).

Closer to our times, I think of Malvina Reynolds's song-poem 'Little Boxes'. 'Ticky-tacky' houses that 'all look just the same', boys who routinely 'go into business / and marry and raise a family' – what a powerful piece to expose the conformity trends sweeping the United States in the 1960s, and even more visible today. And do you remember Reynolds's 'What Have They Done to the Rain?', written to protest the nuclear testing that was putting strontium 90 in the atmosphere?

Today's environmental poetry differs from nature-themed poetry of the past as a result of the growing realization of the extent of human involvement. The dominant motif in today's nature poems is the notion that in everything that happens, humans and nature are and have been actively engaged and inextricably wound. Poetry that reflects this change, this acknowledgement of what is being called the epoch of the Anthropocene, is thriving today. Increasingly it is called 'ecopoetry' and its writers referred to as 'ecopoets'.

Human consciousness, though, is not necessarily front and center in ecopoetry. Often humans in eco-poems speak with the voices and thoughts of other entities, of plants and animals, rocks and Gaia, to express the emotions of fear, or loss, sadness, or shame for all of damaged Creation. In 'Tigerless', written by a youth, Rachel Lewis, winner of a Cape Farewell poetry contest, the speaker is a male tiger searching for his missing father who finds only 'men / And the carcasses of lands they leave behind'. The poem ends with poignant lines that one cannot help but connect to the author and by extension to our impotent fearful selves. The young tiger says that he fears for his lost, unreachable father, 'But Father, I am still more afraid for me' (Lewis 2014).

Expect, when reading ecopoetry, to be made uncomfortable, even angry at times. A poem by John Latham, a research scientist and for many years president of the International Commission on Atmospheric Energy, imagines human babies as defendants in a future 'Judgement at the Court of Gaia' (Latham 2008: 83). Their lawyer urges mercy on the grounds that the children are victims whose parents learned from their mistakes. Gaia replies in words that challenge our comforting view of an all-forgiving nature. Granting that the infants are guilt-free, Gaia nevertheless feels disgust at human failure to acknowledge the tragic loss of the whole panoply of creatures large and small who have evolved in diverse ways to fit and magnify their habitats with beautiful adaptations such as whale talk and dawn songs, lives forever thoughtlessly destroyed. She declares the planet 'can no longer afford the human race' and sentences the babies to death with yet a purpose. '*Feed them to the wolves!*'

Time enters these eco-poems, not always as harsh judgement or the frantic 'so little time', but as a hoped-for return to something approaching pre-lapsarian harmony. Gary Snyder's oft-quoted 'stay together / learn the flowers / go light' (Snyder 1974) is from 'For the Children', a poem that envisions 'valleys, pastures' where 'we can meet' 'in peace' 'if we make it'. Similarly, poet and geographer Eric Magrane's 'Velvet mesquite' speaks from the deeply buried roots of a desert tree that feels a 'pulse' from 'another world', one known to mastodons and sloths and the native Hohokam. The roots declare that in some future time when humans have again learned to live with their fellow creatures 'its pulse will course through me' (Magrane 2016: 51).

III.

I believe that we can heartily second the observation of poet Alison Hawthorne Deming that 'communicating information about climate change has hardly brought the forces of greed, guns, and gutting of the planet to their knees'. 'Information', she adds, 'doesn't change people ... Sometimes passion changes people. Sometimes empathy does' (Deming 2014: 207).

Empathy and passion are characteristic of the poetry of climate change. Were it possible I would include in full a powerful poem written by Pepper Trail, an Ashland, Oregon poet and biologist who works for the United States Fish and Wildlife Service. Its title: 'Carbon Dioxide Reaches Highest Level in Human Existence' (Trail 2014). Trail expresses his outrage at the climate-change deniers and 'incorporated people' who have allowed the carbon dioxide level to rise to 400 parts per million. He laments the futility of young persons who have marched with 350.org banners for what is suddenly not possible. The poem is a howl, a dirge, a tantrum, a cry for lost beauty, lost meaning, for a beloved world about to disappear forever.

But before the poem ends, the speaker cannot help but add, as any heart-sore caring individual cannot help but add: 'So what to do with this reflex of hope? / The twitch of resistance, the irrepressible involuntary outrage / all the ways of caring about the damage done ...'. The speaker has no answer.

We as climate-change educators, as teachers and scientists, as health-professionals, lawyers and ethicists need to lace our lectures with strong poetry, now more than ever. The public needs to feel our anger as well as digest our facts. At a time when lies are widely and casually disseminated, passion and empathy are a form of litmus paper.

As for rage – it is totally appropriate. 'Rage, rage against the dying of the light', Dylan Thomas beseeched his mortally ill father. Climate change is a dying of the light, not alone for individual human beings but for all living things. It is the consequence of a massive act of enclosure, what is left after other interests have cordoned off our commons and privatized the very air we breathe, our health, our food, our plants and animals. Global climate change has stolen our birthright. This insight is behind Vancouver poet Stephen Collis's 'On Blackberries and the Poetic Commons' (Collis 2014), an essay that builds on the way that blackberries invade the edges of private property and are a 'shared community resource'. In a separate poem 'Blackberries' Collis takes words by Thoreau, whom Emerson designated 'captain of a huckleberry party', and arranges them five to a line as blackberries are arranged in nature (Collis 2005). It is an example of how ecopoetry weaves nature and humans together and how at times the resulting tapestry is strange, like a Picasso painting, birds and banks, shrubs and nuthatches displayed side by side in startling arrangements (ibid.).

IV.

Climate change and its consequences are unstoppable in the foreseeable future. Say it.

And yet. When I look into the eyes of our grandchildren and their friends, and think of their children and their children's children condemned to an artificial world, never to know the joys of nature that I have known; when I listen to the squawking woodpeckers in our mesquite tree who don't know they are on their way out, I am unable to be comforted by the exceedingly slim possibility of long-range restitution. I feel that twitch of resistance that Pepper Trail calls a 'reflex of hope'. Eco-poetry is full of such twitches and the most hopeful focus on the joy and power of communal action. Ross Gay's 'To the Fig Tree on 9th and Christian', my new favourite poem, is one example (Gay 2015). It begins with its protagonist rambling absentmindedly along lonely city byways, unobservant, alone in his thoughts, and ends after he has helped an old woman clean a street of the slippery fruit of a fig tree, in the process becoming part of a neighbourly crowd whose participants feed each other from the tree's ripe fruit in an impromptu banquet: 'strangers maybe, / never again'.

This emphasis on communal action is a change from the stress on individual responsibility that anchors much of environmental literature. While individual responsibility is assuredly worthwhile, it can be over-valued and self-congratulatory. The British poet Paul Munden mocks it in 'Mitigation' by naming actions frequently cited as desirable counters to climate change and exposing their ego-building consequences – switching to low wattage bulbs and feeling 'enlightened', for instance, or taking credit for watering the lawn with rainwater. He says, 'I recycle my newspapers / and the headlines go away' (Munden 2008: 57).

Yet in the new ecopoetry communal gatherings do not necessarily escape mockery either. In 'One World Down the Drain' Simon Rae ironically suggests 'another conference' to deal with changing weather patterns (Rae 2007).

So what to do with this reflex of hope? Indigenous author Allison Adelle Hedge Coke suggests in the prose poem 'We Were in a World' that because 'the People' have 'gone away from oratory, song, [and] oraliteratures', they are 'into gyrations reeling' (Coke 2014). One solution is to get back to oratory and song, the province of poetry.

'Sometimes passion changes people. Sometimes empathy does.' I want to end by suggesting that you read, or read again, one of the most passionate, empathic poems I know. Better yet, you can listen to its author recite it. She is a young mother, Kathy Jetnil-Kijiner of the Marshall Islands, and her poem, 'dear matafele peinam' was chosen for the Opening Ceremony of the United Nations Climate Summit in September, 2014 (Jetnil-Kijiner 2014). She begins by addressing her seven-month-old daughter, who is, she says 'so excited for bananas, hugs and / our morning walks past the lagoon'.

It is the lagoon and its predicted rise due to the heating-up of the ocean's waters, inundating their island home and eventually forcing the mother and child's removal from the island, that occasions the poem. Jetmil-Kijiner refuses to accept that this will happen. She models strength and defiant hope, citing a multitude of reasons not to believe in the inevitability of the sea's rise including the action of 'thousands' in active protest 'chanting for change NOW'; citizens petitioning in myriad creative ways involving art, music, and writing; canoes blocking coal ships; the radiance of solar villages. And although her admonition to her daughter to 'just close those eyes, baby / and sleep in peace' 'because we won't let you down' 'you'll see' seems further than ever from realization, it is a moving cry of hope that addresses what is undeniably our greatest threat.

We who would educate the public to the need to respond to climate change NOW need poets and poetry in part because we live in a world where other verbal forms of persuasion are suspect, a world where 'fake news' is not an oxymoron, where scientists are accused of falsifying data, and a climate-change denier gets to head the Environmental Protection Agency. The American poet Alison Hawthorne Deming asks, 'Hasn't language been used to manipulate, oppress, deceive, and betray more consistently than it has ever served as a vessel for aesthetic or spiritual feeling?' (Deming 2014: 206). Yet poetry more often than not gets a free pass because it is widely perceived as such a vessel, writing that is on the side of the reader.

The gag order put into effect by the present US administration that limits public statements on climate change by research scientists is unbelievably bad, yet we in science and public service fields have our own censorious protocol for emotional content in our journals. Rage is edited out, although in order to save Earth we desperately need not just the facts but ways to convey our anger at how

Earth has been treated, as Pepper Trail's poem does so well. We who seek solutions to climate change in adaptation need to understand the fear of even greater loss, as Rachel Lewis's 'Tigerless' does, and need also to acknowledge the ancient emotional connection of humans and nature as Eric Magrane does, and the grief that Mary Oliver expresses for lost beauty. We who are in danger of just giving up and letting come what may, desperately need, in addition to graphs and equations and ethical formulations, poetic expressions of hope in communal action to address climate change such as that of Kathy Jetnil-Kijiner.

Acknowledgements

In 2015 I took part in a seminar, 'The Poetry of Climate Change', led by geographer and poet Eric Magrane. This course coincided with a year-long reading and lecture series at the University of Arizona Poetry Center attended by hundreds of public citizens. The course and the poets who read attempted to answer the question, 'What role does poetry have in anticipating, shaping – or even creating – our future?' and are the basis for these thoughts.

References

Alter, A. (2017) 'American Poets, Refusing to Go Gentle, Rage against the Right', *The New York Times*, 22 April, ppA1, A11.

Boyle, J. (2003) 'The Second Enclosure Movement and the Construction of the Public Domain', *Law and Contemporary Problems*, vol 33.

Clare, J. (2005) 'The Mores', in G. Summerfield (ed.), *John Clare: Selected Poems*, Penguin Classics, London.

Coke, A. A. H. (2014) 'We Were in a World', in *Streaming*, Coffee House Press, Minneapolis, MN

Collis, S. (2005) *Blackberries*, BookThug, Vancouver, Canada.

Collis, S. (2014) 'Of Blackberries and the Poetic Commons', *The Commons*, 2nd edn, Talonbooks, Vancouver, Canada.

Deming, A. H. (2014) *Zoologies: On Animals and the Human Spirit*, Milkweed Editions, Minneapolis, MN.

Felsteiner, J. (2009) *Can Poetry Save the Earth? A Field Guide to Nature Poems*, Yale University Press, New Haven, CT.

Gay, R. (2015) 'To the Fig Tree on 9th and Christian', in *Catalog of Unabashed Gratitude*, University of Pittsburgh Press, Pittsburgh, PA.

Hirshfield, J. (2013) 'Global Warming', in D. Cummings (ed.), *So Little Time: Words and Images for a World in Climate Crisis*, Green Writers Press, Brattleboro VT.

Jetnil-Kijiner, K. (2014) 'Dear Matafele Peinam', UN Climate Summit Poem, available at www.youtube.com/watch?v=DJuRjy9k7GA (accessed 17 August 2017).

Kilmer, J. (1913) 'Trees', in *Poetry*, August, p160.

Latham, J. (2008) 'Judgement at the Court of Gaia', in P. Munden (ed.), *Feeling the Pressure: Poetry and Science of Climate Change*, British Council, Berne, Switzerland.

Lewis, R. (2014) 'Tigerless', Cape Farewell/ Poetry Society, London.

Magrane, E. (2016) 'Velvet Mesquite', in E. Magrane and C. Cokinos (eds), *The Sonoran Desert: A Literary Field Guide*, University of Arizona Press, Tucson, AZ.

McEwen, C. (2008) 'In Memorium', in P. Munden (ed.), *Feeling the Pressure: Poetry and Science of Climate Change*, British Council, Berne, Switzerland.

Merwin, W. S. (1988) 'Witness', in *The Rain in the Trees*, Knopf, New York, NY

Munden, P. (2008) 'Mitigation', in P. Munden (ed.), *Feeling the Pressure: Poetry and Science of Climate Change*, British Council, Berne, Switzerland.

Oliver, M. (2008) 'From This River, When I Was a Child, I Used to Drink', in *Red Bird: Poems by Mary Oliver*, Beacon Press, Boston, MA.

Rae, S. (2007) 'One World Down the Drain', in N. Astley (ed.), *Earth Shattering: Ecopoems*, Bloodaxe Books, Hexham.

Snyder, G. (1974) 'For the Children', in *Turtle Island*, New Directions, New York.

Trail, P. (2014) 'Carbon Dioxide Reaches Highest Level in Human Existence' in *ISLE: Interdisciplinary Studies in Literature and Environment*, vol 21, no 1, pp 135–136.

Wilbur, R. (2004) 'Advice to a Prophet' in *Collected Poems 1943-2004*, Harcourt, New York.

21 Can the Earth Charter movement be renewed?

The covenantal promise of the Earth Charter movement

J. Ronald Engel

Can the Earth Charter movement be renewed?
Need it be?
Should it be?
How, on what terms, and by whom?

The hope and the reality

At its launch in 2000 our hopes for the Earth Charter were high indeed.

We hoped that it would serve as an inspiration and guide for a movement with power to inaugurate a new epoch in world history, variously referred to as the 'great transition' or 'great work' – or as Naomi Klein and our Canadian friends now call it, 'the Leap' – when global ethics would play a material role in world affairs.

We hoped that it would soon become widely recognized as a major step forward in the evolution of international soft law, a Peoples' Charter that would catalyse a movement for justice, sustainability and peace throughout international civil society as well as among the governments of the nations of the world.

The promise of the Earth Charter is unabashedly millenarian. The text explicitly declares:

> "As never before in history, a common destiny beckons us to seek a new beginning." The implication is that a new era of cosmic harmony is imminent. Principle 16 anticipates that harmony: "… peace is the wholeness created by right relationships with oneself, other persons, other cultures, other life, Earth, and the larger whole of which all are a part."
>
> (Corcoran, Vilela and Roerink 2005)

In retrospect, the Earth Charter was a product of a remarkable period in world history when hopes for a time when humanity was at peace with itself and all of creation ran high on many fronts. I do not want to minimize the substantial contributions members of the drafting committee made to the Earth Charter text. Nor would the Charter have been possible without the strong intellectual and personal leadership Steven Rockefeller provided to the consultative and drafting process. Nonetheless, I must say that for the most part the Earth Charter wrote itself. We were the scribes recording the remarkable strides the liberal international community and grassroots activists were then making to set forth a holistic vision of world community founded on a deep reverence for life. The Earth Charter is largely a summarization of what numerous civil society draft 'Earth Charter' texts expressed in the run up to the Rio de Janeiro Earth Summit in 1992 together with a careful review of the progressive history of soft and hard international environmental law since the founding of the United Nations. This is readily evident in the architecture of the Earth Charter, built as it is on the four pillars that have come to define the widely embraced overarching global ethic of 'sustainability' (Engel 2014a):

- respect and care for the community of life;
- ecological integrity;
- social and economic justice; and
- democracy, nonviolence and peace.

Who would have anticipated that seventeen years after its launch amid great fanfare at the Peace Palace in The Hague we would be where we are today? Far from entering a new axial age of justice, sustainability, and peace with all creation the new millennium has revealed how far we have travelled into the Anthropocene and the apocalyptic prospect of an end of planetary life.

And where is the Earth Charter movement in the twenty-first century? It has much to its credit. Over seven thousand organizations have formally endorsed the Charter. The Earth Charter Initiative, headquartered at the University of Peace in Costa Rica, has faithfully served as the nerve centre for the movement, sponsoring a website and innovative programs in education, most prominently through its partnership with the UN Decade of Education for Sustainable Development (2005–2014). The Charter has been the subject of numerous academic conferences and publications, including those by members of the Global Ecological Integrity Group, and has served as the foundation for groundbreaking proposals for the ecological reconstitution of international environmental law. In 2015 the Earth Charter was referenced by Pope Francis in his historic call for a universal 'ecological conversion' to care and respect for our 'common home'.

Yet there is no denying the fact that after the Johannesburg Summit in 2002 the Charter has been largely missing from the public stage. It did not become the popular rallying cry of any of the prominent pro-democracy, pro-environment, pro-peace protest movements of the new century, from the protests in cities across the world against the invasion of Iraq by the United States in 2003, to Occupy Wall Street, Tahrir Square, Taskin Square, to the massive People's Climate March of 2014, right up to the 2015 Paris Climate Conference where it apparently played no explicit role. Current political trends across the planet are unravelling our hope for a new age of ethically-driven international cooperation in the new millennium.

Is the time of the Earth Charter over?

A covenant with Earth

I have argued for some years now that only an explicit, principled, lasting covenant with Earth – democratically conceived, initiated and implemented in new structures of local and global governance –will enable us to successfully make the great leap across the Anthropocene.

I have further held that the Earth Charter is *primus inter pares* among the numerous attempts by civil society to articulate the substance of such a covenant for our epoch (Engel 2002, 2004, 2007, 2008, 2011). A covenant with Earth is the new beginning the Earth Charter announces as the promise of the new millennium. I have not been alone in this interpretation.

In 2001 a group of artists in Vermont built what they called an 'ark of hope' modelled on the treaty practices of the ancient Middle East, an elaborately painted wooden chest bearing a papyrus on which they hand wrote the full text of the Earth Charter. They carried this new 'ark of the covenant' to the United Nations headquarters in New York, and then to the World Summit on Sustainable Development in Johannesburg the following year.

Public intellectuals such as Terry Tempest Williams have also spoken of the Earth Charter as a new covenant with the Earth:

> I like to think of the Earth Charter as a promise to our planet. We will be present with you. We will not turn our backs on the poverty of our brothers and sisters. We will not walk away from our commitment to face whatever the future may bring. We will watch, listen, speak, and act on behalf of a 'reverence for life'.
>
> (Williams 2008)

The only way the Earth Charter vision will become a reality is by a vigorous and explicit embrace of the covenantal faith of Earth Democracy. Earth Democracy conceived as faithfulness to the evolutionary life of the planet and the historical struggle of the human species to live in freedom, equality and solidarity is a wholly different way of living in the world than the one that currently prevails under the name of 'democracy' and is accepted as 'normal' in our contemporary cultures.

Without being named as such, the Earth Charter has the rhetorical structure of a classic covenant: it opens with an assertion of our original belonging to the universal community of life, confesses our alienation from that community, and then lays out the moral imperatives we must follow if we are to be reconciled with the fundamental source of our being and fulfil the promise of our creation. Covenant defines a community of people who pledge to be faithful to one another in their shared faithfulness to a cause of ultimate importance, goodness and truth. In the case of the Earth Charter, the great cause of ultimate importance, goodness, and truth is no less than the salvation of the Earth, and the community of people who pledge their mutual support and accountability to one another in shared commitment to this cause make up what is vital and powerful in the Earth Charter movement.

From where I stand, the failure of the Earth Charter movement to achieve the traction in world affairs we anticipated is due in no small measure to our failure to take the covenantal character of the document with sufficient seriousness. Insofar as this is true, the renewal of the movement depends upon rectifying these failures.

Let me share four ways I believe this can happen.

The renewal of the Earth Charter movement will come when it renews its leadership for an ongoing critical dialogue on global ethics

The first step in renewing the Earth Charter movement is for the participants in the movement to once again exercise leadership for the global ethics dialogue that gave the Charter birth, a dialogue which necessarily requires ongoing criticism and debate about the adequacy of the Earth Charter text itself.

The drafting committee never assumed the Charter was complete and perfect. Loyalty to the Charter could only grow, not diminish, if we took it seriously enough to continue to debate its adequacy, making amendments when justified, and treating it as a living rather than a dead document.

Covenantal relationships require constant critical reflection on the adequacy of the purposes of the covenant as well as the success or failure of the partners to keep it. They are the opposite of a once and for all creedal orthodoxy, or a contract whose terms are limited and set.

Indeed, the Earth Charter closes with the explicit admonition: 'We must deepen and expand the global dialogue that generated the Earth Charter, for we have much to learn from the ongoing collaborative search for truth and wisdom.'

I have previously called attention to the fact that 'equality', with the exception of gender relationships, is absent from the text of the Charter. We need a strong understanding of substantive economic and political equality, equality of condition as well as equality of rights, in the Earth Charter (Engel 2014b). The Earth Charter fails to adequately address a number of such foundational values and issues that are at the centre of our contemporary debates in global ethics and governance. A few examples follow.

To the disappointment of many environmental ethicists and activists, at the last minute in the Charter's drafting, the affirmation of the 'intrinsic value' of every form of life was changed simply to 'value' because the Dalai Lama objected to that term on technical philosophical grounds. Neither does the Charter mention or affirm what is often judged to be the ethical corollary of 'intrinsic value' which is the idea of the 'rights of nature', or the rights of sentient animals. With respect to the treatment of animals generally, the drafting committee was only able to overcome the differences between representatives of the Circumpolar North who are dependent upon hunting for their food and representatives of Eastern faiths that place a high emphasis on compassion for all living beings by finessing the wording so that 'compassion' appears in foundational Principle 2 that deals with our attitude toward life as a whole but not in Principle 15 that specifically addresses the treatment of animals.

Population and reproduction are issues which the Charter similarly avoids addressing head-on. While the Charter's Preamble states that 'An unprecedented rise in human population has overburdened ecological and social systems', and Principle 7 asserts 'Adopt patterns of production, consumption, and reproduction that safeguard Earth's regenerative capacities, human rights, and community well-being', it does not call for a reduction in world population growth. In my judgment our capacity to ethically self-govern our reproductive capacities is the acid test of our ability to ethically self-govern ourselves in all other respects.

Surely the issue of climate change needs to be explicitly flagged and addressed in the Earth Charter text! And with it the critical concepts of 'planetary boundaries'. And in light of the claim of the United States to a 'right to pre-emptive aggression' as well as the whole complex of issues involving national and territorial sovereignty and the 'right of self-determination' in numerous conflicts across the world today Principle 16 on peace and security cries out for much further development.

There is ample opportunity for the Earth Charter to not only update its own ethical vision with the help of a renewed global ethics dialogue, but to take leadership in moving that dialogue forward in these troubled times.

Since the launch of the Charter in 2000 many other outstanding declarations of global ethics have come on the scene, including the Universal Declaration on the Rights of Mother Earth, the dozen or so draft citizen treaties prepared for the Rio+20 Summit in 2012, the Declaration of the Rights of Indigenous Peoples, and the Oslo Manifesto for Ecological Law and Governance. Insofar as the Earth Charter still occupies the position of primus inter pares of global ethics declarations, the most comprehensive statement of the Earth covenant in the making, there has long been the opportunity for the Earth Charter Initiative to convene a summit of the leadership of the most significant of these declarations. A primary task of such a summit would be to examine the synergies as well as differences with the Sustainable Development Goals (SDGs) currently being used to frame all international dialogue and negotiations concerning environment and development. Such a summit would go far to reignite the dialogue on global ethics and found an international covenanted alliance with greater organizational power and moral influence than any one of them can exercise independently.

Of equal importance is the opportunity for the Earth Charter leadership to challenge every community throughout the world to draft an 'ethical declaration' or 'local earth charter' detailing its specific ethical responsibilities in light of the needs and strengths of its particular biocultural region, and in the context of the special contributions it can make to our shared task of implementing the universal principles of the international Earth Charter across the planet.

A covenanted world is federal, simultaneously local, national, and international. As Hannah Arendt observed in the wake of the adoption of the Universal Declaration of Human Rights, the universal 'rights of man' require actual polities to guarantee them. It is a matter of 'common but differentiated responsibilities', based on the recognition that the planetary community of life and the human polities that govern it are a community of communities, and hence the global polity which should ideally govern the whole should be founded on a covenant of covenants. Unfortunately the Earth Charter does not explicitly reference or provide a clear definition of the principle of 'common but differentiated responsibilities' which is further reason for renewing the global ethics dialogue.

Leonardo Boff and Mirian Vilela make a good case for how recent social movements in Brazil have been motivated by an ethical vision that in principle is very similar to the Earth Charter (Boff and Vilela 2014). Naomi Klein's Canadian 'Leap Manifesto' is an example of an ethical manifesto that is also very compatible with the Earth Charter (Klein 2017). Do we not now need to take the next step and explicitly connect such transformative national initiatives with the Earth Charter itself so that the Earth Charter movement may be clearly understood as a matter of 'common but differentiated responsibilities', a federal, global covenanted movement, involving commitments of universal and local responsibility by every community and nation across the globe?

The renewal of the Earth Charter movement will come when it takes a strong stand on behalf of the truth of its covenantal vision and principles

The Earth Charter movement can be renewed when it is willing and able to argue that its vision and principles are rationally and empirically justified truths and not merely because they enjoy international consensus. We have confused justification with legitimation.

The Earth Charter text makes strong truth claims throughout – scientific, political, moral and spiritual – beginning at the outset with precisely the kind of ontological truth claim that grounds the covenantal worldview. The Earth Charter asserts that we and all other beings are uniquely valuable members of the community of life and this 'fact' entails 'oughts': certain universal ethical responsibilities, rights and duties, that are incumbent upon us to honour as morally self-conscious human members of the Earth community.

Philosopher of religion Douglas Sturm puts it this way:

> We are, in our fundamental reality, relational beings, co-creators of an evolving universe, bestowed by our inheritance with the special powers of humankind, and therefore held responsible, so far as we are capable, for the flourishing of the entire community of life – in part, at least, for the sake of our own flourishing. That is our calling and, we might say, that is our appointed destiny.
>
> (Sturm 2000)

The Earth Charter announces a prophetic covenant whose moral imperatives stand in judgment of the world and do so on the authority of its claims to truth. Primary among these is the truth that we live or die, outwardly and inwardly, by what we hold to be the cause of ultimate importance worthy of the devotion of our lives. This means that the cause of all causes – the sine qua non of all our many and different hopes for human redemption – is faithfulness to the planet and every human and other being that lives upon it.

The task of the Earth Charter movement is to bear witness to the ways in which living in covenant with the Earth is synonymous with what Vaclav Havel called 'living in the truth'.

Moral justification is more than a matter of abstract metaphysical or empirical argument however. It is also, and for most people primarily, a persuasive and coherent narrative that makes sense of the passage of time in its immediate personal meaning, and gives us a place and a role to play in the drama of the cosmos. The sacred story we need to tell to ground the Earth Charter as a revelation of the truth of our cosmic and historical unfolding is once again fundamentally covenantal: human history is a matter of the making, breaking, reforming and renewing of covenants; it is a struggle to make the covenants of our civilization more responsive to the life-sustaining relationships of our existence, more inclusive in their membership, more respectful and caring of the integrities of their members, more holistic in their grasp of the multiple moral concerns that must be met for communities to thrive. Yet humans also have the capacity (as no other creature appears to have) of deceit and deliberate betrayal of the covenants of our being – betrayal of the universal covenant of life into which we are born, and betrayal of the covenants to which we have voluntarily sworn allegiance, and these betrayals are central to our sacred story as well. Indeed, the betrayal of the Earth is the primary plotline which human civilization as a whole is now living out.

The story of the universe as told by Thomas Berry has also been popular among Earth Charter proponents and insofar as it locates the Earth at the centre of the cosmic drama today, it surely provides inspiring insights into the kind of historical responsibilities we have as a species. This story needs to be supplemented now by the story of how the covenants of Earth Democracy have been apprehended, incarnated and betrayed throughout the course of world history. This wider and deeper natural and historical sacred story is the necessary context for appreciating the power of the story Earth Charter proponents often tell as a kind of sacred narrative: the story that begins with the Charter of the United Nations and then moves step by step to increasingly inclusive declarations of universal ethical principle to the Earth Charter. The high calling of the Earth Charter movement is to bear witness to this story in its full cosmic and historical sweep and in all its light and dark reality.

The renewal of the Earth Charter movement will come when it takes clear well-reasoned positions on the great issues of our age, and stands in solidarity with those who are taking action

To be in covenant means to be true to one's word and stand in solidarity with those who are doing likewise. The text of the Earth Charter is far too abstract to gain substantial popular understanding and support unless it can be shown to lead to well-reasoned positions regarding the policies that are ethically mandated to resolve the most critical of the social, political and economic issues facing the world today, and visible support for concrete initiatives that embody them. Senior adviser to Earth Charter International Brendan Mackey has recently argued:

> A casual reader of the Earth Charter, giving it a quick scan, might find it an inspiring text but one that gives little direction by way of implementation, as it does not specify on who or where the many responsibilities implied by its 77 principles lay, and is soft on enforcement given that the document lacks any accountability measures. A more careful reading, however, reveals that while the text may lack such teeth it does contain a host of imperatives demanding strong if not radical action.
>
> (Mackey in press)

Arguments for the actions that the Earth Charter requires if it is to be implemented are being worked out now by Mackey and other leading scholars and activists in the fields of global ecology, international law, and climate ethics, as well as in particular national political contexts (Bosselmann and Engel 2010; Mackey and Rogers 2015; Bosselman, Brown and Mackey 2012; Javier and Juarez 2014). It remains however for representative spokespersons for the Earth Charter movement to endorse these positions and press them in every available political and academic fora as the positions required to make the Earth Charter a living reality. To take such positions requires moral courage, political and economic independence, and the willingness to risk retribution from powerful interests.

It follows that the Earth Charter movement will not be renewed until it is willing to face and name the fact of our betrayal of the Earth, the agents and agencies that are promulgating it, and the economic, social and political systems that support it, and then work to mobilize the various forms of hard and soft power necessary to transform betrayal into new covenantal commitment. I fully expected that when the United States on false premises violated international law and invaded Iraq in 2003 the Earth Charter Initiative could not fail to condemn it. Neither, to my knowledge, has it taken visible notice of the life and death struggle for indigenous rights taking place only miles from the Earth Charter headquarters at the University of Peace. We do not find notice on the Earth Charter website, for example, of the murder of the internationally recognized indigenous activist, Berta Cáceres, in nearby Honduras.

Perhaps the single greatest failure of the Earth Charter movement is the lack of an accurate and persuasive account of the actual powers and ideologies responsible for the ecocide injustice and violence sweeping the planet or any clear call to the mobilization of non-violent resistance against them. It does not seem to countenance the possibility that people will betray life on purpose, for profit, for the power it gives them, sometimes, like Faust, out of pure resentment that they are mortal.

Does the movement assume corporate capitalist-driven globalization will be the context in which the world will continue to produce, trade and consume for years to come and that sustainability can be achieved without reorganization of the global economy? Without, in other words, a 'new beginning', 'as never before in human history'?

Obviously, no civil society organization can be expected to pass judgment on the myriad ways the many nations, corporations, armies and wealthy elites of the world breach Earth Charter principles. Yet we desperately need paradigmatic examples that reveal the larger patterns of global oppression, and well researched analyses of the underlying power structures driving these patterns. Moreover, as Naomi Klein argues, we must go well beyond resistance to build an international political coalition with sufficient power to actually take the reins of government in each of the nations of the world.

The renewal of the Earth Charter movement will come when it is reconstituted as a covenanted democratic movement

The Earth Charter movement is not organized as a self-governing association in which citizens of the world participate as equal members with specified rights and duties. Instead, the leadership of the movement, seated in the Earth Charter Initiative, is governed by the self-perpetuating board of the Earth Charter Council with no accountability to any institutional entity outside itself.

The Earth Charter movement as a covenanted community assumes a participatory democratic understanding of the governance of the Earth Charter movement itself, one in which each participant when pledging commitment to the Earth Charter vision and principles is also pledging accountability to every other member of the movement in free, equal participatory governance of the movement itself.

The Earth Charter text assumes such a polity for the movement. Earth Charter Principle 3 reads: 'Build democratic societies that are just, participatory, sustainable, and peaceful'.

Underpinning the Earth Charter and its precursors is the fundamental assumption that in the new era of global governance we are seeking to create it is 'we, the peoples of the world' who are sovereign citizens and who bear the moral obligation for personal, political, economic and ecological self-government that is free, equal, communal and sustainable. It is we, the citizens of the world, in our local communities and associations, and in the coalitions we make with one another across the planet, who make the Earth covenant.

We have the challenge therefore of founding a membership organization, or organization of organizations, that will enable the citizens of the Earth Charter movement to take ownership of the Charter. Am I too optimistic to believe the peoples of the world want to be active participants in the authorship and implementation of the covenant that governs our planetary home?

References

Boff, Leonardo and Mirian Vilela. 2014. 'Society, Changes and Social Movements: the Case of Brazil'. In *The Earth Charter, Ecological Integrity and Social Movements*. Edited by Laura Westra and Mirian Vilela. London: Earthscan.

Bosselman, Klaus, P. G. Brown and B. Mackey. 2012. ' Enabling a Flourishing Earth: Challenges for the Green Economy, Opportunities for Global Governance'. *Review of European Community & International Environmental Law* 21: 23–39.

Bosselmann, Klaus and J. Ronald Engel, eds. 2010. *The Earth Charter: A Framework for Global Governance*. Leiden: KIT Publishers.

Corcoran, Peter Blaze, Mirian Vilela and Alide Roerink. 2005. *The Earth Charter in Action: Toward a Sustainable World*. Amsterdam: KIT

Engel, J. Ronald. 2002. 'The Earth Charter as a New Covenant for Democracy'. In *Just Ecological Integrity: the Ethics of Maintaining Planetary Life*, pp. 37–52. Edited by Peter Miller and Laura Westra. Lanham, Maryland: Rowman and Littlefield.

Engel, J. Ronald. 2004. 'A Covenant Model for Global Ethics'. *Worldviews: Religion, Culture, Ethics* 8(1): 29–46.

Engel, J. Ronald. 2007. 'A Covenant of Covenants: A Federal Vision of Global Governance for the 21st Century'. Chapter 2 in *Sustaining Life on Earth: Environmental and Human Health Through Global Governance*. Edited by Colin Soskolne. Lanham, Maryland: Lexington Books.

Engel, J. Ronald. 2008. 'The Covenantal Foundations of Governance for the Community of Life'. Chapter 2 in *Governance for Sustainability: Issues, Challenges, Cases*. Edited by Klaus Bosselmann, Ron Engel and Prue Taylor. Bonn, Germany: Earthprint.

Engel, J. Ronald. 2011. 'Property: Faustian Pact or New Covenant with Earth?' In *Property Rights and Sustainability: The Evolution of Property Rights to Meet Ecological Challenges*, pp. 63–86. Edited by David Grinlinton and Prue Taylor. Leiden and Boston: Martinus Nijhoff Publishers.

Engel, J. Ronald. 2014a. ' Sustainability'. In *Encyclopedia of Bioethics*, 4th edn, pp. 3040–3051. Editor-in-Chief, Bruce Jennings. New York: Macmillan.

Engel, J. Ronald. 2014b. 'Prologue: Summons to a New Axial Age – the Promise, Limits, and Future of the Earth Charter'. In *The Earth Charter, Ecological Integrity and Social Movements*. Edited by Laura Westra and Mirian Vilela. London: Earthscan.

Javier, Francisco, and Camarena Juarez. 2014. 'The Earth Charter. An Instrument of Environmental Policy in Mexico-a Soft Law or Hard Policy Perspective. In *The Earth Charter, Ecological Integrity, and Social Movements*. Edited by Laura Westra and Mirian Vilela. London: Earthscan.

Klein, Naomi. 2017. *No is Not Enough: Resisting Trump's Shock Politics and Winning the World We Need*. Chicago, IL: Haymarket Books.

Mackey, Brendan. 2018. http://earthcharter.org/virtual-library2/earth-charter-relevance-contemporary-issues/ 'A Reflection on The Earth Charter Project and its Mission in the Anthropocene'.

Mackey, B. and N. Rogers. 2015. 'Climate Justice and the Distribution of Rights to Emit Carbon'. Chapter 13 in *Access to International Justice*. Edited by P. Keyzer, V. Popovski and C. Sampford. New York: Routledge.

Sturm, Douglas. 2000. 'Identity and Alterity: Summons to a New Axial Age'. Available at www.earthcharterinaction.org/invent/images/uploads/2000_sturm_1_2.pdf (accessed 9 May 2014).

Williams, Terry Tempest. 2008. 'Taking the Globe to Our Bosom'. In *A Voice for Earth*, p. xvii. Edited by Peter Blaze Corcoran and A. James Wohlpart. Athens, GA: University of Georgia Press.

Conclusion

The ever-increasing importance of ecological integrity in international and national law

Klaus Bosselmann

Introduction

The success of twenty-five years of the Global Ecological Integrity Group (GEIG) is materially owed to Laura Westra and her extensive work on ecological integrity. In many books (for example, Westra 1994, 1998, 2016), she has highlighted the connections between science, ethics and law to describe the multi-disciplinary, multi-faceted concept of ecological integrity – an 'umbrella' concept as she calls it.

In science, integrity describes to the functioning, robustness and resilience of systems (systems integrity). In ethics, integrity refers to truthfulness, moral responsibility and consistency of character. To this end, it is – or should be – fundamental to the professional work of scientists, lawyers, politicians and decision-makers (personal integrity). In law, integrity relates to the interpretation of law as comprising rules and principles as well as socio-political and ethical components (law as integrity; see Dworkin 1986). Furthermore, human rights have the protection of human dignity (physical and mental integrity) at their core.

Despite the significantly different trajectories, science, ethics and law share an understanding of integrity as a concept of wholeness, coherence and cohesion.

From this perspective, integrity reveals itself as a multi-disciplinary, foundational concept. With the attribute 'ecological' it becomes even more foundational. The ecological system is the most encompassing system we know. Without the protection of Earth's ecological systems, all human development (incl. socio-economic systems) is unsustainable, literally without foundation.

GEIG's numerous conferences and books, together with the work and advocacy of its members have created a wealth of information, an ongoing source of inspiration and – crucially – a key and tool for overcoming the global ecological crisis. As observed in the introduction to this book, GEIG has developed in three phases: clarification (until 2000), dissemination and implementation. The 'implementation stage' is to promote ecological integrity through law and governance, the subject of this chapter.

Most human actions and interactions are guided by law (of whatever content and scope). The law can therefore have a critical role for steering society into the right (i.e. sustainable direction). Codifying and institutionalizing sustainability is the key challenge of our time. Sustainability law and governance may not be everything, but everything else is nothing without it.

As will be shown here, ecological integrity has been increasingly recognized international and domestic environmental law precisely for the reasons mentioned. Making the connections between ethics, science and law is of great importance. They help us to understand that these disciplines share a certain level of normative guidance. If defined as integrity, these disciplines pursue – or should pursue – wholeness, coherence and cohesion. The integrity concept therefore provides guidance for seeking coherence of social systems and environmental systems. This is the point.

For too long, the dichotomy between human and natural spheres – replicated in the divide between science and humanities, law and ethics, economics and the environment – has been the stumbling stone for understanding 'what's really going on'. What's really going on is that Earth is a coherent whole with life (in all its forms) being part of it. Earth is not just a backdrop for human enterprise. We can't live without it and – bar leaving this planet – need to reconcile our needs with the needs of Earth.

It is appropriate therefore to begin this survey with a look at how environmental jurisprudence has changed over the recent 20 or so years. Earth as an ecological whole, not states, has become a central reference point. This change of thinking makes ecological integrity so relevant. As we will then see, international and domestic law has begun to adopt the integrity of Earth's ecological system as an overarching objective of law and a duty for states.

The Earth-centred approach in environmental law

New concepts such as Earth systems law (Bosselmann 1995), Earth jurisprudence (Burdon 2011), Earth justice (Cullinan 2003), Earth governance (Bosselmann 2015a), Earth democracy (Shiva 2015) or Earth as a legal entity (Magalhães et al. 2016) consider Earth as a central reference point for the design of environmental law. This puts them potentially at odds with state-centered 'international' environmental law and indeed the very concept of 'environmental' law. These and other ethically motivated concepts have generated a rich body of literature on ecological approaches to environmental law (Bosselmann and Taylor 2017). The many proponents of ecological law from all around the world have recently established the Ecological Law and Governance Association (www.elga.world).

Earth is more than the sum of its parts. Earth is a unifying entity in her own right and very different from the world (of nations). Earth is one and borderless, the world is divided and diverse. In legal terms, the world has international law resembling division and diversity, while the Earth has – nothing. There is as yet nothing in the law responding to Earth's wholeness and complexity. The standard explanation for this void is that Earth is not a legal category. The world is made up of people and countries represented through states. Hence we have international law, but not global law or Earth law in a true sense.

Seen through the lenses of states, the wholeness of Earth remains obscure and politically irrelevant. Correspondingly, planetary systems such as the global commons (climate, biodiversity, oceans etc.) appear as discrete entities and removed from the 'real' world. Yet, what could be more real than the recognition of one undivided Earth? There is a mismatch between the international law of states and the reality of one Earth.

The alternative is not a law without states, but a law informed by ecological realities. Earth's ecological systems are not there to serve humans needs, they are simply there. It is this recognition of reality that is currently missing in our international and national laws. So our predicament is straightforward: unless law recognizes and internalizes ecological realities, it is doomed to fail. No amount of rhetoric ('green economy', 'sustainable development', 'sustainable development goals') can gloss over the simple truth that humans are utterly dependent on the integrity of Earth's ecological systems and need to govern themselves accordingly.

The project of internalizing ecological realities is of course not new. It is as old as the history of environmental law. The very purpose and subject of environmental law (i.e. protection of the natural environment) should have led to a fundamental 'greening' of law and governance. Instead, the opposite happened. Since their early beginnings in the 1960's environmental laws have aimed for less (i.e. 'balancing' environmental, social and economic interests). As if there was a middle ground between life and death, environmental laws and policies assumed a position of looking after ecological systems while leaving the growth paradigm untouched. However, there cannot be a compromise between sustaining life and life-threatening, cancerous growth. Under the growth paradigm environmental protection was bound to remain at the political periphery rendering environmental laws to temper around the edges (Bosselmann 1995, 2010; Gaines 2014).

Perhaps surprisingly, such obvious design flaws have never occupied the discipline of environmental law in any major way. Overlooking mainstream scholarship of 40 or so years, there has been a relative dearth of critical literature. Most environmental lawyers tend to see themselves foremost as lawyers rather than as ecologists (if at all). However, viewing ecological realities from the traditional legal perspective, is likely to result into anthropocentrism and reductionism. Viewing political and legal realities from an ecological perspective, on the other hand, is likely to result into ecocentrism and holism (Bosselmann 1995). From its beginnings environmental law scholarship has been divided along this clash of paradigms.

Yet, we have all good reasons to believe that the paradigm clash will be resolved in favour of ecocentrism as we have one great ally on our side (i.e. Mother Earth). There is now overwhelming evidence that Earth systems are no longer functioning as they must to sustain human life. We are seeing the signs of planetary decline all around us: the oceans, the atmosphere, the biosphere are systems in crisis and so are human systems: countries and communities are increasingly impoverished socially, economically and environmentally. With the exception of 1 per cent (or less) of humanity, the entire world population of human and non-human beings is suffering. It is as if humanity has given up on herself. Like Dennis Meadows, co-author of *Limits of Growth* (Meadows et al. 1972), said 30 years later:

> I have tried long enough to be a global evangelist and have learned that in doing so that I cannot change the world. Apart from that, humankind behaves like someone committing suicide and there is no point in arguing with a suicidal person once they have already jumped out of the window.
> (Meadows 2012)

While we cannot be certain about suicidal behaviour or a life-saving turnaround, we can describe a rescue strategy. At the very least, such knowledge will be useful for attempts to rebuild civilization following eventual collapse. Central to any rescue strategy must be the objective to not overstep planetary boundaries and to protect the integrity of ecological systems. In legal terms, this objective can be expressed as a fundamental principle or *grundnorm* (Kim and Bosselmann 2015) underpinning the law in general (not just environmental law) and in a similar way as respect of human dignity or equality, fairness and justice. To get there, two initial steps are necessary.

The first step is recognizing the reality of planetary boundaries. Of the nine boundaries identified thus far (Rockström et al. 2009), three have already been exceeded (atmospheric greenhouse gas concentrations, rate of biodiversity loss and nitrogen cycle) (Steffen et al. 2015). The recognition of planetary boundaries sets a non-negotiable bottom-line for all human activities. More particularly, and in the context of the well-trodden concept of sustainable development, it suggests a hierarchical order of its three constituent elements: the natural environment is fundamental and comes first, human social organization exists within it and comes second and economic modelling only exists within both, neither in parallel nor above them. Only such a hierarchical understanding of sustainable development ('strong sustainability') reflects the reality of planetary boundaries and marks the first step towards a refined rule of law.

The second step is taking the strong sustainability approach to the design and interpretation of laws. One of the most basic legal concepts in this regard is the rule of law. It is the prime tool to ensure control and accountability of governments and implies that all citizens are subject to the law. But not any law can rely on the rule of law in order to be respected. Rather, the rule of law has a moral dimension. It could not be any other way. Historically, the rule of law developed alongside autonomy for the individual, equality and social justice. This makes it inconceivable, for example, to consider a law justifying the killing of all blue-eyed, blond people as in any way consistent with the rule of law. Such a law would be invalid not because of its immorality, but because it was not enacted in accordance with the procedures authorized by the rule of law. Likewise, we could conceive of an interpretation of the rule of law that invalidates any law threatening humanity's natural living conditions. Positively speaking, respect for planetary boundaries and the integrity of Earth's ecological systems are non-negotiable preconditions for the rule of law. I have called this the rule of law grounded in the Earth (Bosselmann 2013).

Ecological Integrity as a *grundnorm*

The never-ending discourse around sustainable development – including sustainable development goals – has yet to discover the fundamental importance of sustainability – a concept as old as humanity and well practised in most societies until the arrival of the industrial age (Bosselmann 2015b). This also true for Europe. In fact, the origins of the modern concept of ecological sustainability are in forest management theory of the seventeenth century.

During this era the notion of 'sustainability' was created to describe the need of ongoing, future-oriented care for forests, the main energy source of the time. The term itself was coined by German engineer and forest economist Hans Carl von Carlowitz who analysed the European resource crisis while crossing the continent between 1665 and 1669. Inspired by his studies of the British and French literature as well as by the experience he gained in his administrative function, Carlowitz eventually wrote his magnum opus *Sylvicultura oeconomica oder Naturmaessige Anweisung zur Wilden Baum-Zucht* ('Forest Economy or Guide to Tree Cultivation Conforming with Nature'). The book, published in 1713, focused on 'how such conservation and growing of timber can be managed in order to provide continued, durable and sustained use'. This marked the birth of the German term *nachhaltend*, later translated to 'sustained' and 'sustainable' in English.

Philosophers like Spinoza, Leibniz, Schelling, Goethe, Herder and Hegel promoted an idealistic, holistic and organic worldview, and did not regard the model of sustainability as limited to forest management (Bosselmann 2017). Rather, sustainability reflected an appropriate interpretation of life as a whole. Cotta's first essay of 1792 expressed this universal attitude: 'In the whole world there is no thing without relationship to something else'. That also reflects the creation of the term 'Oecologie' (from Greek 'oikos' for 'house', 'household') by Ernst Haeckel, in 1866, in allusion to philosopher Johann Gottfried Herder. Previously, Herder characterized the Earth as our 'Wohnplatz' ('living space', 'home') and the human role as 'haushalten' ('housekeeping').

About the same time, forest academies in Austria-Hungary, Switzerland, France, Russia, Scandinavia, the United Kingdom with its colonies and finally the United States started to adopt the German prototype of sustainable forestry. Consequently, the German word *Nachhaltigkeit* was in need of suitable translation. In 1837, Professor Adolphe Parade, the director of the French Forest Academy in Nancy, translated the word with 'production soutenu' revealing its Latin roots as 'sustinere' (from 'tenere' to keep). Like the English expression 'sustainable', this resembles the notions of enduring, lasting, keeping up, maintaining, carrying on, continuing and sustaining (Spindler 2012).

Yet, during the subsequent era of the Industrial Revolution, the principle of 'sustainability' became increasingly irreconcilable with the idea of linear progress and indefinitely growing wealth that gained momentum around 1800. Starting in England and finding its way from Western Europe to America, one of the most important changes in the history of humankind began. It was stimulated by philosophical streams like Newton's model of physics and the mechanistic-atomistic image of exploitation of natural resources and largely overlooked consideration of ecological sustainability. Fossil fuels (coal, oil) substituted renewable energy sources like wood and wind. Finally, natural boundaries had to be stretched to match the needs of the demographic change. Corresponding to this development, the idea of local, public and common responsibility was rejected to give way to a 'private free enterprise' approach with its premise of granting individuals absolute dispositional powers over their properties, including environmental goods. As the relation to land became increasingly marked by self-serving, competing neighbour rights rather than by ecological concerns, the individual even raised itself above nature (Spindler 2012).

However, the development during the Industrial Revolution did not render the idea of sustainability redundant. With regard to present-day's global crisis concerning ecological, economic and social structures in equal measure, it is now of greater topicality than ever before. By the early 1980s sustainability was clearly understood as the anti-thesis of the growth paradigm. Its core meaning is the preservation of the integrity of ecological systems (Bosselmann 2017). There is ample evidence that the Brundtland Commission aimed for sustainable development as ecologically sustainable development and as a counter model to the 'more-is-better' growth paradigm of modern economies. Powerful corporates and states prevented sustainable development from becoming such a counter model. Instead, sustainable development remained the meaningless concept of mashing environmental, social and economic interests.

The preservation of the integrity of Earth's ecological system has been featuring in modern environmental law for a long time.

It has been incorporated in many domestic conservation laws (Bosselmann 2017) and international environmental law where it first appeared in the 1974 Great Lakes Water Quality Agreement between Canada and the United States. Since then some 25 international environmental treaties and agreements refer to ecological integrity as a general objective as, for example, the preamble and Article 7 of the 1992 Rio Declaration on Environment and Development ('States shall cooperate in a spirit of global partnership to conserve, protect and restore the health and integrity of the Earth's ecosystem') or Article 40 of the Rio+20 outcome document The Future We Want ('We call for holistic and integrated approaches to sustainable development that will guide humanity to live in harmony with nature and lead to efforts to restore health and integrity of the Earth's ecosystem'). The 2000 Earth Charter is in its entirety designed around the concept of ecological integrity. For example, principle 5 urges 'all individuals, organizations, businesses, governments, and transnational institution' to '[p]rotect and restore the integrity of Earth's ecological systems, with special concern for biological diversity and the natural processes that sustain life'. Similarly, Article 2 of the 2010 IUCN Draft International Covenant on

Environment and Development states: 'Nature as a whole and all life forms warrant respect and are to be safeguarded. The integrity of the Earth's ecological systems shall be maintained and where necessary restored.' This inclusion is significant because the Draft Covenant is a codification of existing environmental law and a blueprint for an international framework convention.

The same is true for the 2017 Global Pact for the Environment, initiated by the French Government and intended to become a UN treaty. Its Preamble refers to 'respecting the balance and integrity of the Earth's ecosystem'. And Article 2 stipulates that 'every state or international institution, every person, natural or legal, public or private, has the duty to take care of the environment' defined as contributing 'to the conservation, protection and restoration of the integrity of Earth's ecosystem.'

Other examples include the 2015 Paris Agreement on Climate Change with its Preamble ('Noting the importance of ensuring the integrity of all ecosystems'), Article 4.13 ('promote environmental integrity') and Article 6.2 ('ensure environmental integrity') or the 2016 World Declaration on the Environmental Rule of Law with its Preamble ('Life depends on the integrity of the biosphere and the interdependence of ecological systems' and 'The environmental rule of law should serve as the legal foundation for achieving environmental justice, global ecological integrity and a sustainable future for all.') and Principle 4 ('Legal and other measures shall be taken to protect and restore ecosystem integrity and to sustain and enhance the resilience of social-ecological systems').

Applying the usual standards for the recognition of concepts as international law, it would be possible to say that the repeated and consistent references to ecological integrity amount to an emerging fundamental objective or *grundnorm* of international environmental law.

A fundamental rule to not harm the integrity of Earth's ecological or planetary systems would be a significant step towards ecological law and a reflection of the need to respect our planetary boundaries. As Michelle Maloney puts it in her comprehensive Earth jurisprudential analysis:

> Ecological integrity is both the 'end game' and the starting point. It is the end game because humanity needs to ensure human societies do not interrupt the ecological integrity of ecosystems in their region and nations. It is also the starting point for any regulatory regime for living within our limits we must put the overall health of the Earth first, and fit the human project within its frame.
>
> (Maloney 2014)

Conclusion

GEIG and its members have made a considerable contribution to the development of environmental law. First its critique and second its transformation towards the ecologicalization of law in general. Ecological law and governance is no longer a distant dream, but an emergent characteristic of jurisprudence and law for our time. This time or epoch has been referred to as the Anthropocene. The Anthropocene must not be misunderstood as an acknowledgment of humanity's ultimate triumph over nature and an invitation to keep going in this way. Only cynics would think so. To the contrary, the notion of the Anthropocene is a timely wake-up call for realizing the fact that humanity has, to its own peril, overstepped Earth's ecological boundaries. Nothing could be more urgent now than respecting these boundaries and create a safe operating space for humanity.

It is here, where the concept of ecological integrity comes in. If we only appreciate its fundamentality to life on Earth, then *Homo sapiens* should be able to apply it to his/her own benefit. Ecological integrity must guide law and governance in the same vein as human rights have guided law and governance prior to the discovery of our survival crisis.

References

Bosselmann, K. (1995) *When Two Worlds Collide: Society and Ecology*, RSVP, Auckland

Bosselmann, K. (2010) Losing the Forest for the Trees: Environmental Reductionism in the Law. *Environmental Laws and Sustainability*, Special Issue of *Sustainability* 2(8): 2424–2448.

Bosselmann, K. (2013) The Rule of Law Grounded in the Earth, in L. Westra and M. Vilela (eds), *The Earth Charter, Ecological Integrity and Social Movements*, Routledge, New York, 3–11.

Bosselmann, K. (2015a) *Earth Governance: Trusteeship of the Global Commons*, Edward Elgar Publ. Cheltenham.

Bosselmann, K. (2015b) Sustainability and the Law, in P. Salmon and D. Grinlinton (eds), *Environmental Law in New Zealand*, Thomson Reuters, Wellington, 73–103.

Bosselmann, K. (2017) *The Principle of Sustainability: Transforming Law and Governance*, 2nd edn, Routledge, New York.

Bosselmann, K. and Taylor, P. (eds) (2017) *Ecological Approaches to Environmental Law*, Edward Elgar Publ., Cheltenham.

Burdon, P. (ed.) (2011) *Exploring Wild Law: The Philosophy of Earth Jurisprudence*, Wakefield Press, Adelaide.

Cullinan, C. (2003) *Wild Law: A Manifesto for Earth Justice*, Green Books, Totnes.

Dworkin, R. (1986) *Law's Empire*, Fontana Press, London.

Gaines, S. (2014) Reimaging Environmental Law for the 21st Century, *Environmental Law Reporter* 44(3): 10188.

Kim, R. and Bosselmann, K. (2015) Operationalizing Sustainable Development: Ecological Integrity as a *Grundnorm* in International Law, *Review of European Community and International Environmental Law* 24(2).

Magalhães, P., Steffen, W., Bosselmann, K., Aragao, A. and Soromenho-Marques, V. (eds) (2017) *SOS Treaty – Safe Operating Treaty: Managing Earth Systems Use*, Cambridge Scholars Publ., Cambridge.

Maloney, M. (2014) The Role of Regulation in Reducing Consumption by Individuals and Households in Industrialised Nations, *PhD thesis*, Griffith University, Brisbane, Australia.

Meadows, D. (2012) Humanity is Still on the Way to Destroying Itself. *Der Spiegel*, 7 December, www.spiegel.de/international/world/limits-to-growth-author-dennis-meadows-says-that-crisis-is-approaching-a-871570.html.

Meadows, D. H., Meadows, D. L., Randers, J. and Behrens, W. (1972) *Limits to Growth*, Potomac Ass., Hanover.

Rockström, J. (2009) Planetary Boundaries: Exploring the Safe Operating Space for Humanity, *Ecology and Society* 14(2): 32.

Shiva, V. (2015) *Earth Democracy: Justice, Sustainability, and Peace*, new ed., North Atlantic Books, Berkeley, CA.

Spindler, E. (2012) Geschichte der Nachhaltigkeit. Vom Werden und Wirken eines beliebten Begriffes, in J. Jenkings and R. Schröder (eds), *Sustainability in Tourism. A Multidisciplinary Approach*, Gabler, Wiesbaden.

Steffen, W. (2015) Planetary Boundaries: Guiding Human Development on a Changing Planet, *Science* (347): 791.

Westra, L. (1994) *An Environmental Proposal: The Principle of Integrity*, Rowman & Littlefield Publ., Lanham, MD.

Westra, L. (1998) *Living in Integrity: A Global Ethic to Restore a Fragmented Earth*, Rowman & Littlefield Publ., Lanham, MD.

Westra, L. (2016) *Ecological Integrity and Global Governance: Science, Ethics and Law*, Routledge, New York.

Index

www.ingramcontent.com/pod-product-compliance
Ingram Content Group UK Ltd.
Pitfield, Milton Keynes, MK11 3LW, UK
UKHW011000180926
475701UK00004B/480
* 9 7 8 0 3 6 7 5 1 0 9 0 9 *